T0174119

Small Scale
OPTICS

Small Scale
OPTICS

Preecha Yupapin ▪ Jalil Ali

CRC Press
Taylor & Francis Group
Boca Raton London New York

CRC Press is an imprint of the
Taylor & Francis Group, an **informa** business

CRC Press
Taylor & Francis Group
6000 Broken Sound Parkway NW, Suite 300
Boca Raton, FL 33487-2742

First issued in paperback 2019

© 2014 by Taylor & Francis Group, LLC
CRC Press is an imprint of Taylor & Francis Group, an Informa business

No claim to original U.S. Government works

ISBN-13: 978-1-4665-9234-6 (hbk)
ISBN-13: 978-0-367-37952-0 (pbk)

This book contains information obtained from authentic and highly regarded sources. Reasonable efforts have been made to publish reliable data and information, but the author and publisher cannot assume responsibility for the validity of all materials or the consequences of their use. The authors and publishers have attempted to trace the copyright holders of all material reproduced in this publication and apologize to copyright holders if permission to publish in this form has not been obtained. If any copyright material has not been acknowledged please write and let us know so we may rectify in any future reprint.

Except as permitted under U.S. Copyright Law, no part of this book may be reprinted, reproduced, transmitted, or utilized in any form by any electronic, mechanical, or other means, now known or hereafter invented, including photocopying, microfilming, and recording, or in any information storage or retrieval system, without written permission from the publishers.

For permission to photocopy or use material electronically from this work, please access www.copyright. com (http://www.copyright.com/) or contact the Copyright Clearance Center, Inc. (CCC), 222 Rosewood Drive, Danvers, MA 01923, 978-750-8400. CCC is a not-for-profit organization that provides licenses and registration for a variety of users. For organizations that have been granted a photocopy license by the CCC, a separate system of payment has been arranged.

Trademark Notice: Product or corporate names may be trademarks or registered trademarks, and are used only for identification and explanation without intent to infringe.

Library of Congress Cataloging-in-Publication Data

Yupapin, Preecha P.
  Small scale optics / Preecha Yupapin, Jalil Ali.
      pages cm
  Includes bibliographical references and index.
  ISBN 978-1-4665-9234-6 (hardcover : alk. paper)
  1. Photonics. 2. Integrated optics. I. Ali, Jalil. II. Title.

TA1520.Y87 2014
621.36′93--dc23   2013025841

Visit the Taylor & Francis Web site at
http://www.taylorandfrancis.com

and the CRC Press Web site at
http://www.crcpress.com

Contents

Preface

The behavior of light in small scale optics or nano/micro optical devices has shown promising results, which can be useful for any fundamental and applied research, especially in nanoelectronics. In this book, a new design for a small scale optical device, in particular, a microring resonator device, is proposed; a design that can be used to generate forms of light on a chip, where optical spin, antenna, and whispering gallery mode are the major applications. Most of the chapters use the proposed device, which is made up of silica and InGaAsP/InP with a linear optical add-drop filter incorporating two nonlinear micro/nano rings on both sides of the center ring (add-drop filter). This particular configuration is known as a "PANDA" ring resonator. Light pulse, for instance, Gaussian, and bright and dark solitons are fed into the system through different ports such as an add port and through port. By using the practical device parameters, the simulation results are obtained using the Optiwave and MATLAB™ programs. Results obtained by both analytical and numerical methods show that many applications can be exploited. In applications, when the practical device parameters are used, then such a device can be fabricated and implemented in the near future.

The proposed system can be used for many applications, especially when the device is coated with a metallic material. The applications such as atom/molecule transportation, molecule/atom antenna, atomic/molecular automata, atom/molecule electronic devices, cells/atom distributed sensors, cells/atom radio, cells communications, micro-Faraday cage, molecular/atom storage and logistics, atom/molecule trapping (storage) or cooling, everlasting atom/molecule investigation, atomic/molecular storage for quantum gate or computer application, storm search and navigator sensors, magnetic/spin networks, magnetic net, microplasma source, 3-D flat panel device, and large cooling area (volume) can be constructed in the same way as a single device. This book presents the use of the optical nonlinear behaviors for spins, antenna, and whispering gallery mode within micro/nano devices and circuits, which can be used to perform in many applications such as nano-antenna, molecular filter and motor, magnetic therapy, radiotherapy, spin transport, cell communications, fast and slow light, optical network sensors, cancer treatment by short pulse laser, and in future challenges. The theoretical background of the basic PANDA ring resonator device is also given.

P. P. Yupapin and J. Ali

MATLAB™ is a registered trademark of The MathWorks, Inc. For product information, please contact

The MathWorks, Inc.
3 Apple Hill Drive
Natick, MA 01760-2098 USA
Tel: 508-647-7000
Fax: 508-647-7001
E-mail: info@mathworks.com
Web: www.mathworks.com

Authors

Preecha P. Yupapin, Ph.D., received his Ph.D. in electrical engineering from the City University of London in 1993. He was a postdoctorate research fellow in 1994 under the European Community research project. Dr. Yupapin has been working with the Department of Physics, Faculty of Science, King Mongkut's Institute of Technology, Ladkrabang, Bangkok, Thailand since 1985. Since 2007, he has been a visiting professor with the Department of Physics, Faculty of Science, Universiti Teknologi, Malaysia. Dr. Yupapin has authored and coau-

thored more than 550 research papers in the Google Scholar database, 35 papers in Pubmed, and 35 chapters and books. His research interests are in nanophysics, nanoelectronics, spintronics, nanocommunication and networks, molecular electronics, nanomedicine and beauty, nanoenergy, quantum information, and human engineering. Dr. Yupapin is a member of the Thai Institute Physics (TIP), South East Asia Theoretical Physics Association (SEATPA) committee, South East Asia Collaboration for Ocean Renewable Energy (SEAcORE), and the Optical Society of America (OSA). He has had several outstanding works, one of which was related to the Nobel Prize Awards in Physics in 2012 on the "Stopping Light" concept in *IEEE–Photonics Technology Letters*, 2009. He was a candidate of the Eni Awards (Nobel Prize in Energy) in 2013 on the "New Type Solar Cells" in *Optics Express*, 2012, and recently a work on "Whispering Gallery Modes (WGMs) of light in a PANDA Ring Waveguide" has been accepted to publish in *Nature: Scientific Reports* in 2013.

Jalil Ali, Ph.D., received his Ph.D. in plasma physics from Universiti Teknologi Malaysia (UTM) in 1990. At present, he is a professor of photonics at the Institute of Advanced Photonics Science, ESci Nano Research Alliance, and the Physics Department of UTM. He has authored and coauthored more than 400 technical papers published in international journals, five books, and a number of book chapters. Dr. Ali's areas of interest are in FBGs, optical solitons, fiber couplers, and

nanowaveguides. He is currently the head of the Nanophotonics Research Group, ESci Nano Research Alliance, UTM. Dr. Jalil Ali is a member of OSA, SPIE, and the Malaysian Institute of Physics.

1 Nonlinear PANDA Ring

1.1 INTRODUCTION

The majority of the optical bistable devices have been widely investigated since their practical usage in optical gate systems [1,2], optical switching [3], optical memory [4], optical flip-flops [5–7], and photonic signal processing [8], where switching is performed in the optical domain [9]. Some research on fiber ring resonators has been done to obtain optical bistable devices [10–12]. The majority of the analysis on the optical bistable behavior of fiber systems has been carried out using the so-called graphical method. The graphical method was developed by Felber et al. [13,14] to study the dynamic transmission properties of a nonlinear fiber ring resonator. Felber also examined the optical bistability of the optical Fabry–Perot system containing a Kerr medium [15]. Ikeda [16], Agrawal [17], and Nakatsuka [18] have applied an iterative method and linear stability analysis [16] in the initial examination of the ring system. All-optical switching devices based on ordinary optical fiber offer great promise for reducing the complexity of optical systems for nonlinearity and they are responsible for the optical bistability. The PANDA Vernier resonator has superiority in that no pumping components are needed, the size of the PANDA resonator system is microsized, and relative output/input power consumption is quite low. Additionally, our proposed system can also be used for ultrafast all-optical switch communication systems [19]. One of the approaches to analyze the complex photonic circuits and fast calculation of optical transfer functions in the Z-domain is the signal flow graph (SFG) method, proposed by Mason [20,21] and developed by Moslehi et al. [22]. There are two practical approaches to highlight the nonlinear effects: increase the length of the fiber, or increase the optical power within the fiber [12]. One of the approaches to achieve nonlinearity in ring resonators with fixed radius is an iterative method whereby the input optical pulse should rotate iteratively through a ring resonator to compensate the required long length for emergence of nonlinear phenomena [10,23–27]. To attain bistability in the ring resonator, the idea of using the Vernier effect for inserting two small ring resonators to a common add/drop filter, which is named the *PANDA ring resonator* system (Figure 1.1a), is proposed. Lateral small ring resonators in the PANDA resonator system are used to generate nonlinearity in an optical feedback process. Hence, lateral small rings in the PANDA Vernier resonator system play the role of long length fiber loops. The PANDA Vernier resonator can be used as an ultrafast all-optical switching (AOS) device, which is necessary in high-bit-rate communication systems [19].

In this chapter, the graphical approach with an SFG in the Z-domain is used to drive the optical transfer function for through and drop port of the PANDA Vernier

1

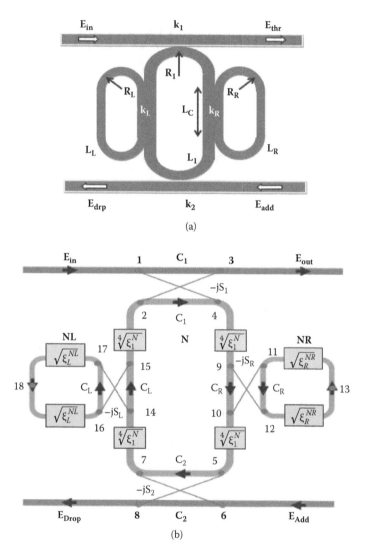

FIGURE 1.1 Architecture of PANDA Vernier filters. (a) A waveguide layout. (b) A Z-transform diagram SFG.

resonator, which is fabricated from racetrack silicon-on-insulator (SOI) waveguides. The dependence of through and drop port powers and the transmission characteristics of the silicon PANDA Vernier filter on the coupling coefficients of symmetric nondirectional couplers and different resonant mode numbers are studied. The relative through/input and drop/input port power consumption, insertion loss, and switching time for two resonant mode numbers with different coupling coefficients are determined to achieve the highest efficiency with the lowest relative output/input power consumption. The proposed PANDA Vernier system can be used as a proper component for fast shift keying (FSK) applications [28–30] and optical signal processing [3,8] in photonics integrated circuits.

1.2 OPTICAL TRANSFER FUNCTION

The schematic of an optical modified add-drop resonator with two added small ring resonators (we call this system a *PANDA Vernier resonator*) is demonstrated in Figure 1.1, which is assembled by 2×2 optical couplers. The 2×2 direct optical couplers can be represented in an SFG diagram. For each coupler, the fraction of light passes through two photonic nodes in the same path (Figure 1.1b) expressed by $C_i = \sqrt{(1-\gamma)(1-k_i)}$ where k_i is the coupling coefficient for the ith coupler (i = 1, 2, R, L) and γ is the intensity loss for each coupler. In contrast, the fraction of light passed through the cross path including two photonic nodes from different paths is expressed by $S_i = \sqrt{(1-\gamma)k_i}$. The Z-transform parameter is stated as $Z^{-1} = \exp(-jk n_{eff} L_i)$, where $k = 2\pi/\lambda$ is the vacuum wave number, n_{eff} is the effective refractive index of the waveguide, and $L_i = 2\pi R_i + 2L_c$ is the circumference of each racetrack ring, where R_i is the radius of each ring (R_1 for main ring, and R_R and R_L for lateral right and left small rings, respectively) and the L_c shows the straight sections of the racetrack resonators (Figure 1.1a). The free spectral range (FSR) of the device is determined by $FSR_i = c/n_g L_i$, where $n_g = n_{eff} + f_0(dn_{eff} / df)_{f_0}$ is the group refractive index of the ring, n_{eff} is the effective refractive index, and f_0 is the design (center) frequency [17]. The SFG for a PANDA Vernier resonator is shown in Figure 1.1b in which $E_1 = E_{in}$, $E_3 = E_{thr}$, and $E_8 = E_{drp}$ are considered as the input node, through node, and drop node, respectively. The FSR of the symmetric PANDA Vernier resonator consists of a main racetrack ring with length of $L_1 = 2\pi R_1 + 2L_c$ and two identical small rings with length of $L_R = 2\pi R_R + 2L_c$ and $L_L = 2\pi R_L + 2L_c$, expressed as

$$FSR_{tot} = N.FSR_1 = NR.FSR_R = NL.FSR_L \tag{1.1}$$

Here, N is the resonant number for the main ring resonator of the PANDA resonator system, and NR and NL are order mode numbers for the right and left small lateral rings of a PANDA resonator, respectively. All of these resonant numbers are coprime and integers (NR and NL < N).

The optical transfer function or input–output transmittance relationship from input node $E_i(z)$ to output node $E_n(z)$ in an SFG can be determined by Mason's rule [20,21,30]

$$H = \frac{E_n}{E_i} = \sum_{i=1}^{n} \frac{T_i \Delta_i}{\Delta} \tag{1.2}$$

where H is the input and an output port network function, T_i shows the gain of the ith forward path from input to exit port and n is the overall number of onward paths from input to output [31,32]. The SFG determinant displays by Δ, which is given by $\Delta = 1 - \sum_{i=1} L_i + \sum_{i \neq j} L_i L_j - \sum_{i \neq j \neq k} L_i L_j L_k + ...$ Here, the L_i is the transmittance gain of network for the ith loop [33]. The products of nontouching loops are induced in the multiplicative summations. The loops that have no node in common are considered nontouching or separate loops [34,35]. The symbol Δ_i in H is the determinant Δ after all loops, which touch the T_i path at any node are eliminated [36]. To determine Δ, all of the independent and nontouching loops in the system should be determined. As can be seen in Figure 1.1b, six independent loop gains of the SFG can be identified for the PANDA resonator system, which are expressed as (here, the quantities that belong to the right and left small lateral rings of the PANDA resonator system are shown after the R and L indices),

$$L_1 = C_1 \, C_2 \, C_R \, C_L \, \xi^N \tag{1.3}$$

$$L_2 = L_R = C_R \, \xi_R^{NR} \tag{1.4}$$

$$L_3 = L_L = C_L \, \xi_L^{NL} \tag{1.5}$$

$$L_4 = L_{1R} = -C_1 \, C_2 \, C_L \, S_R^2 \, \xi_R^{NR} \, \xi^N \tag{1.6}$$

$$L_5 = L_{1L} = -C_1 \, C_2 \, C_R \, S_L^2 \, \xi_L^{NL} \, \xi^N \tag{1.7}$$

$$L_6 = C_1 \, C_2 \, S_R^2 \, S_L^2 \, \xi_L^{NL} \, \xi_R^{NR} \, \xi^N \tag{1.8}$$

Here, $\xi_i^p = X_i \, Z^{-p}$ are defined as the multiplication of $X_i = \exp(-\alpha L_i/2)$, one round-trip loss coefficient for each of the rings at the PANDA resonator (main, right, and left) to $Z^{-p} = \exp(-j\varphi p)$, the transmission of light along the ring resonator (the closed pass), which is called the *Z-transform parameter*. Here, $\varphi = k \, n_{eff} L$ is phase shift and p is defined as an integer resonant mode number for each ring in the PANDA resonator system. There are also five possible products of transmittances of two nontouching loops, given as

$$L_{12} = L_1 \cdot L_R = C_1 \, C_2 \, C_R^2 \, C_L \, \xi_R^{NR} \, \xi^N \tag{1.9}$$

$$L_{13} = L_1 \cdot L_L = C_1 \, C_2 \, C_L^2 \, C_R \, \xi_L^{NL} \, \xi^N \tag{1.10}$$

$$L_{23} = L_R \cdot L_L = C_L \, C_R \, \xi_L^{NL} \, \xi_R^{NR} \tag{1.11}$$

$$L_{25} = L_R \cdot L_5 = -C_1 \, C_2 \, C_R^2 \, S_L^2 \, \xi_L^{NL} \, \xi_R^{NR} \, \xi^N \tag{1.12}$$

$$L_{34} = L_L \cdot L_4 = -C_1 \, C_2 \, C_L^2 \, S_R^2 \, \xi_L^{NL} \, \xi_R^{NR} \, \xi^N \tag{1.13}$$

and one possible product of transmittances of nontouching loops including main and lateral small rings

$$L_{123} = L_1 \cdot L_2 \cdot L_3 = C_1 \, C_2 \, C_L^2 \, C_R^2 \xi_L^{NL} \, \xi_R^{NR} \, \xi^N \tag{1.14}$$

To determine the optical transfer function for through port of the PANDA Vernier system, all of the onward transmittance routes from input node ($E_1 = E_{in}$) to through port ($E_3 = E_{thr}$) should be specified. Five forward path transmittances from node 1 to node 3 for the through port are recognized such that all of them also touch the loops. Here, forward routes are demonstrated with T_i. The symbol Δ_i in the following equations shows the ith loop determinant, $\Delta_i = 1 - \sum_{i=1} L_i + \sum_{i \neq j} L_i L_j - \sum_{i \neq j \neq k} L_i L_j L_k + ...$, that includes the loops, which do not have any node in common with ith transmittance path. The forward path transmittance for direct path 1→3 is given as

$$T_1^{thr} = C_1$$

$$\Delta_1 = 1 - \sum_{i=1}^{6} L_i + L_{12} + L_{13} + L_{23} + L_{25} + L_{34} - L_{123}$$

(1.15)

The second forward path transmittance for the path through main ring 1→4-9-10-5-7-14-15-2→3 is given as

$$T_2^{thr} = -C_2 C_R C_L S_1^2 \xi^N$$

$$\Delta_2 = 1 - L_2 - L_3 + L_{23}$$

(1.16)

The third forward path transmittance for the path includes main and small right rings 1→4-9-12-13-11-10-5-7-14-15-2→3, which is given as

$$T_3^{thr} = C_2\, C_L S_1^2\, S_R^2 \xi_R^R \varsigma \xi^N$$

$$\Delta_3 = 1 - L_3$$

(1.17)

The fourth forward path transmittance for the path consists of a main ring and two small rings 1→4-9-12-13-11-10-5-7-14-17-18-16-15-2→3, which is given as

$$T_4^{thr} = -C_2\, S_1^2\, S_R^2\, S_L^2\, \xi_R^R \xi_L^L\, \xi^N$$

$$\Delta_4 = 1$$

(1.18)

The fifth forward path transmittance for the path comprised of the main and small left rings 1→4-9-10-5-7-14-17-18-16-15-2→3 is given as

$$T_5^{thr} = C_2\, C_R\, S_1^2\, S_L^2\, \xi_L^L\, \xi^N$$

$$\Delta_5 = 1 - L_2$$

(1.19)

Replacing Equations (1.3) through (1.19) into Equation (1.2), and considering $S_i^2 + C_i^2 = 1$, the optical transfer function for the through port of the PANDA Vernier resonator can be derived as

$$
H_{31} = \frac{E_{Through}}{E_{in}} =
\begin{aligned}
&\left\{ C_1 \left(1 - C_R \, \xi_R^{NR} - C_L \, \xi_L^{NL} + C_R \, C_L \, \xi_R^{NR} \, \xi_L^{NL} \right) + \right. \\
&+ C_2 \, \xi^N \left(C_R \, \xi_L^{NL} + C_L \, \xi_R^{NR} - C_L \, C_R - \xi_L^{NL} \, \xi_R^{NR} \right) \left. \right\} \\
\hline
&\left\{ 1 - C_1 C_2 C_R \, C_L \, \xi^N - C_R \, \xi_R^{NR} - C_L \, \xi_L^L + \right. \\
&+ C_1 \, C_2 \, C_L \, S_R^2 \, \xi_R^{NR} \, \xi^N + C_1 \, C_2 \, C_R \, S_L^2 \, \xi_L^{NL} \, \xi^N \\
&- C_1 \, C_2 \, S_R^2 \, S_L^2 \, \xi_L^{NL} \, \xi_R^{NR} \, \xi^N + C_1 \, C_2 \, C_R^2 \, C_L \\
&\times \xi_R^{NR} \, \xi^N + C_1 \, C_2 \, C_L^2 \, C_R \, \xi_L^{NL} \, \xi^N + C_L \, C_R \\
&\times \xi_L^{NL} \, \xi_R^{NR} - C_1 \, C_2 \, C_R^2 \, S_L^2 \, \xi_L^{NL} \, \xi_R^{NR} \, \xi^N + \\
&- C_1 \, C_2 \, C_L^2 \, S_R^2 \, \xi_L^{NL} \, \xi_R^{NR} \, \xi^N \\
&+ C_1 \, C_2 \, C_L^2 \, C_R^2 \, \xi_L^{NL} \, \xi_R^{NR} \, \xi^N \left. \right\}
\end{aligned}
\tag{1.20}
$$

Here, $\xi^N \equiv X_1 Z^{-N} = e^{-\alpha L_1/2} \cdot e^{-iN\varphi}$, $\xi_R^{NR} \equiv X_R Z^{-NR} = e^{-\alpha L_R/2} \cdot e^{-iNR\varphi}$, and $\xi_L^{NL} \equiv X_L Z^{-NL} = e^{-\alpha L_L/2} \cdot e^{-iNL\varphi}$ are defined as the multiplication of one round-trip loss coefficient to the Z-transform parameter for main, lateral right, and lateral left racetrack resonators, respectively.

To specify, the optical transfer function for the drop port of the PANDA Vernier system loops are the same as determined for the through port but the forward transmittance paths should be identified from input node ($E_1 = E_{in}$) to drop port ($E_8 = E_{drp}$). As demonstrated in Figure 1.1b, two onward transmittance paths can be recognized from node 1 to node 8. The first direction pass via the main ring is the forward path including 1-4-9-10-5→8 nodes

$$
T_1^{drp} = -C_R \, S_1 \, S_2 \, \sqrt{\xi^N}
$$
$$
\Delta_1 = 1 - L_2 - L_3 - L_{23}
\tag{1.21}
$$

The second forward path transmittance, which traverses the main and right small rings for the path comprising 1-4-9-12-13-11-10-5→8 nodes is given by

$$
T_2^{drp} = S_1 \, S_2 \, S_R^2 \xi_R^{NR} \sqrt{\xi^N}
$$
$$
\Delta_2 = 1 - L_3
\tag{1.22}
$$

Substituting Equations (1.3) through (1.14) and Equations (1.21) and (1.22) into Equation (1.2), and considering $S_i^2 + C_i^2 = 1$, the drop port optical transfer function for the PANDA Vernier resonators calculates as

$$H_{81} = \frac{E_{Drop}}{E_{in}} = \frac{\left\{ S_1 S_2 \sqrt{\xi^N} \left(C_R C_L \xi_L^{NL} - C_L \xi_L^{NL} \xi_R^{NR} + \xi_R^{NR} - C_R \right) \right\}}{\left\{ 1 - C_1 C_2 C_R C_L \xi^N - C_R \xi_R^{NR} - C_L \xi_L^L + \right.}$$

$$+ C_1 C_2 C_L S_R^2 \xi_R^{NR} \xi^N + C_1 C_2 C_R S_L^2 \xi_L^{NL} \xi^N$$

$$- C_1 C_2 S_R^2 S_L^2 \xi_L^{NL} \xi_R^{NR} \xi^N + C_1 C_2 C_R^2 C_L \qquad (1.23)$$

$$\times \xi_R^{NR} \xi^N + C_1 C_2 C_L^2 C_R \xi_L^{NL} \xi^N + C_L C_R$$

$$\times \xi_L^{NL} \xi_R^{NR} - C_1 C_2 C_R^2 S_L^2 \xi_L^{NL} \xi_R^{NR} \xi^N +$$

$$- C_1 C_2 C_L^2 S_R^2 \xi_L^{NL} \xi_R^{NR} \xi^N$$

$$\left. + C_1 C_2 C_L^2 C_R^2 \xi_L^{NL} \xi_R^{NR} \xi^N \right\}$$

1.3 RESULTS AND DISCUSSION

The high-index-contrast of silicon-on-insulator enables submicrometer waveguide dimensions as well as the radii of microring resonators to be as small as a few micrometers while ensuring low loss and single mode propagation. In this study, we assume that (1) the silicon-on-insulator waveguides are perfectly rectangular with a height of 220 nm and a width of 500 nm; (2) the group refractive index is wavelength independent; (3) the waveguide curvature is neglected for n_{eff} calculations, and (4) the distance between waveguides is not considered for n_{eff} calculation [37].

We employ the SOI strip waveguides with a width of 420 nm, a height of 220 nm, and a buried oxide (BOX) thickness of 2 μm. We chose two layouts of racetrack ring resonators with different resonant mode numbers. The first configuration is a PANDA resonator system comprised of three racetrack resonators with the straight sections of Lc = 15 μm, and the radii (defined from the centers of the waveguides), R_1 and $R_R = R_L$, of the half-circle sections of the racetrack resonators for the main, lateral small rings are 6.545 μm and 4.225 μm, respectively. The propagation loss is about 20 dB/cm and the group refractive index is 4.5473 [37]. From Equation (1.1), to achieve the overall FSR of 4.65 THz for the PANDA resonator system, the resonant mode numbers should be set on (N;NR;NL) = (5;4;4). For the second layout, we selected three SOI racetrack waveguides with the width of 500 nm and the circumference of 127.91 μm and 99.487 μm [37] for the main and lateral racetrack rings, respectively. The propagation loss for this waveguide is 3 dB/cm and the group refractive index is 4.306 [37]. To achieve the overall FSR of 4.9 THz for the second configuration of the PANDA resonator system, the resonant mode numbers can be determined as (N;NR;NL) = (9;7;7). We also suppose similar intensity insertion loss coefficients γ = 0.01 for all couplers of the silicon PANDA Vernier resonator.

Normally, coupling coefficients of the couplers adjacent to the bus waveguides (K_1 and K_2) are stronger than the innermost couplers. The fixed coupling coefficients of $K_1 = K_2 = 0.50$ are used for the outer couplers. We determine a threshold coupling coefficient for middle couplers of the PANDA Vernier system in that resonant peaks and interstitial peaks for through port insertion loss have an equal value (coupling constant value [κ = 0.001], in Figure 1.2a and Figure 1.3a). This threshold coupling coefficient used as a reference coupling and for two more coupling coefficients bigger and smaller than the threshold coefficient and the optical behavior of the PANDA Vernier system are studied. Since the PANDA Vernier system is symmetric in configuration, we also suppose the symmetric coupling coefficient for

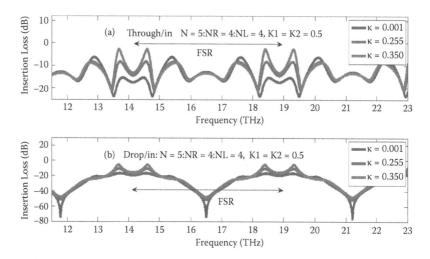

FIGURE 1.2 The influence of variation of middle coupling coefficients on inversion loss of a PANDA Vernier resonator with a resonant mode number of $(N;NR;NL) = (5;4;4)$, $L_c = 15\,\mu m$, $R_1 = 6.545\,\mu m$; $R_R = R_L = 4.225\,\mu m$, $k_1 = k_2 = 0.5$, $\gamma = 0.01$, $\alpha = 20\,(dB/cm)$. (a) Through port response. (b) Drop port response.

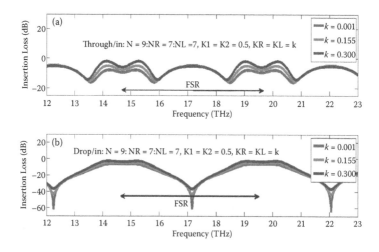

FIGURE 1.3 The effect of changing middle coupling coefficients on inversion loss of a PANDA Vernier resonator with a resonant mode number of $(N;NR;NL) = (9;7;7)$, with a racetrack circumference of $L_1 = 127.91\,\mu m$, $L_R = L_L = 99.487\,\mu m$, $k_1 = k_2 = 0.5$, $\gamma = 0.01$, $\alpha = 3\,(dB/cm)$. (a) Through port response. (b) Drop port response.

a small right and left ring resonator $k_R = k_L = k$. As is shown in Figure 1.2, for the first layout with the resonant mode number of 5;4;4, the symmetric coupling coefficients for couplers between the main ring with lateral small ring resonators are chosen as $k = 0.001$, 0.255 (threshold coupling-value line), and 0.350.

For threshold coupling of 0.225 the on–off ratio for through port (Figure 1.2a) is 12.3 dB. In this coupling coefficient, resonant peaks and interstitial peaks get the same value. For the least coupling coefficient of 0.001, the insertion loss hit 17 dB while the corner resonant peaks suppressed to 9.2 THz. Increasing the coupling coefficient to the value larger than the threshold coefficient causes a trade-off between resonant peaks and interstitial peaks. For the coefficient of 0.350, the insertion loss again hit 17 dB but suppressed resonant peaks for coupling coefficients less than the threshold coefficient were intensified and changed to the resonant peaks for coupling coefficients larger than threshold coupling (for instance k = 0.350). For the drop port of the PANDA Vernier system, the threshold coefficient has the lowest insertion loss of 42.2 dB as can be seen in Figure 1.2b. In this case, increasing the coupling coefficient from 0.001 to 0.35 decreases the insertion loss from 57 dB to 41 dB.

Figure 1.3 shows the insertion loss for through and drop port of the second configuration of the PANDA Vernier system with the resonant mode number of 9;7;7 for the main racetrack perimeter of $L_1 = 127.91$ μm and symmetric small rings with circumference of $L_R = L_L = 99.487$ μm. The procedure is the same as the order mode number of 5;4;4 used for the resonant mode number of 9;7;7. The constant coupling coefficients of $K_1 = K_2 = 0.50$ are employed for the outer couplers and the threshold coupling coefficient of $k_t = 0.155$ recognized for this layout. For two coupling coefficients of k = 0.300 (larger than k_t) and k = 0.001 (smaller than k_t), the optical treatments of the PANDA Vernier system are studied. As is demonstrated in Figure 1.3a, for a threshold coupling of k = 0.155, all resonant peaks and interstitial peaks suppressed equally to 5.2 THz and the on–off ratio reaches 9 dB. For the coupling coefficient of 0.001 and 0.300 the insertion loss hit 7 dB and 12.5 dB, respectively. As can be observed in Figure 1.3a, the resonant peaks interval (13.5–16 THz) are suppressed to 2 THz and 8 THz for k = 0.300 and k = 0.001, respectively. Interstitial resonant peaks between 16–18.5 THz hit 4.8 dB, 5.2 dB, and 5.3 dB for k = 0.001, $k_t = 0.155$, and k = 0.300, respectively. It means that for the through port of the PANDA Vernier system, increasing the coupling coefficient to the value larger than the threshold coefficient causes interchange between resonant peaks and interstitial peaks. As illustrated in Figure 1.3b for the drop port of the PANDA Vernier system, decreasing the middle coupling coefficient contributes to enhancement of insertion loss. For the coupling coefficient of k = 0.300, 0.155, and 0.001, the insertion loss measured 34 dB, 40 dB, and 57 dB, respectively.

Lateral small ring resonators in the PANDA Vernier system are responsible for the nonlinearity in an optical feedback process, instead of using long path fiber or thousands iterating of input pulse in the ring resonator [10,23–27]; hence, lateral small rings compensate long path traveling of the optical pulse to get nonlinearity, consequently the optical bistability can occur in a shorter time.

A bistability diagram of the PANDA Vernier system for variation of relative drop/input power versus variation of through/input power for resonant mode numbers of 5;4;4 is shown in Figure 1.4. For all middle coupling coefficients $k_R = k_L = k$ diagram can be divided into five areas and changing k brings about a shift in the border of these areas. For threshold coupling $k_t = 0.225$, the first point of bifurcation arises at $P_{thr/in} = 0.14$, $P_{drp/in} = 0.01$. Five regions can be separated as follows: the first bistable area $(0.01 < P_{drop} < 0.05)$, the Ikeda-quadratic unstable area $(0.05 < P_{drop/in} < 0.07)$, the Ikeda-quintuple unstable area $(0.07 < P_{drop/in} < 0.10)$, the Ikeda-quadratic unstable area $(0.10 < P_{drop/in} < 0.25)$, and the second bistable area $(0.25 < P_{drop/in} < 0.35)$.

In the bistable regions for each point of relative drop/input power, two points of relative through/input power can be found. In Ikeda-quadratic for every point of relative drop/input power, four points of relative through/input power exist. Multiplicity increases to five points in Ikeda-quintuple unstable areas. The second points of bifurcation take place at $(P_{thr/in} = 0.03, P_{drp/in} = 0.35)$. As is shown in Figure 1.4, variation of the symmetric coupling coefficients causes change in the boundaries of Ikeda and bistable regions.

A bistability diagram of the PANDA Vernier system for variation of relative drop/input power versus variation of through/input power for resonant mode numbers of 9;7;7 with $L_1 = 127.91$ μm and $L_R = L_L = 99.487$ μm is shown in Figure 1.5. The graph can be divided into three regions. For a threshold symmetric coupling coefficient of $k_t = 0.155$, the relative drop/input power region between $0.04 < P_{drop/in} < 0.21$ is considered the first bistable region, $0.21 < P_{drop/in} < 0.38$ is the second area, and $0.38 < P_{drop/in} < 0.39$ is perceived as the third region. The first and third regions are bistable areas. In these areas for every point of drop port power, two points of through port power can be found. As can be observed in Figure 1.5, the first point of bifurcation occurs when the relative through/input power and drop/input power are $P_{thr/in} = 0.04$ and $P_{drp/in} = 0.06$. Ikeda unstable region, $0.21 < P_{drop/in} < 0.38$, goes

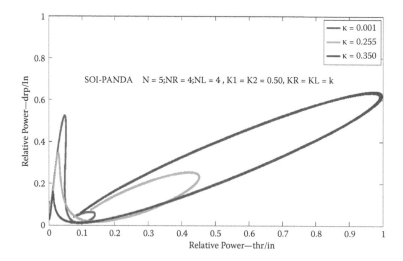

FIGURE 1.4 Optical bistability diagram of a PANDA Vernier resonator for a resonant mode number of 5;4;4.

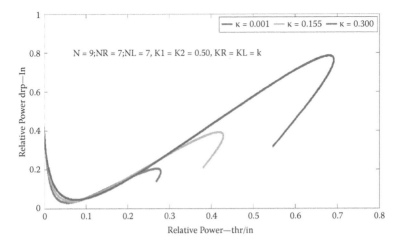

FIGURE 1.5 Optical bistability diagram of a PANDA Vernier resonator for a resonant mode number of 9;7;7.

on the second region, in which each point of relative drop/input power is devoted to three points for through/input power. As can be seen in Figure 1.5, the increasing middle coupling coefficient from $k_R = k_L = k = 0.001$ to 0.300 leads to modification in the boundaries of Ikeda and bistable regions. It is interesting that changing the mode numbers from 5;4;4 to 9;7;7 divides the graph from five regions to three regions.

The optical transfer function behavior for the drop port of the PANDA Vernier resonator with a resonant mode number of 5;4;4 is shown in Figure 1.6a. For a threshold coupling coefficient of $k_t = 0.225$ an interval frequency region less than 1.38 THz, the maximum ratio of 0.64 for drop/input electric field reaches its minimum value of 0.007, which the frequency switching time calculated as 7.2 ps. As is demonstrated in Figure 1.6a, at a specific frequency regime the optical transfer function for drop port reaches plateaus according to the symmetric middle coupling coefficient and the optical transfer function's dependency on symmetric middle coupling coefficients, which can be ignored. Selecting symmetric lateral coupling coefficients of the PANDA Vernier system of 0.001 and 0.350 values leads to calculated switching time of 2.7 ps and 9 ps, respectively.

The optical treatment of the through port's transfer function of the PANDA Vernier resonator with resonant mode number of 5;4;4 is shown in Figure 1.6b. The dependency of through port optical transfer function to symmetric middle coupling coefficients is eliminated after 0.53 THz and the relative through/input port electric field touches to 0.2. The through port switching time for a threshold coupling coefficient of the PANDA Vernier resonator hit 2.1 ps while for coupling coefficients of 0.001 and 0.350 fast frequency switching of 1.1 ps and 3.3 ps were measured, respectively.

Changing the resonant mode numbers to 9;7;7 causes the graph to be sharpened in frequency regime as shown in Figure 1.7. In this resonant mode for threshold coupling

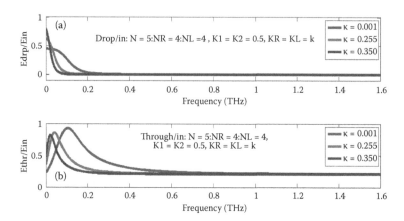

FIGURE 1.6 Optical transfer function of a PANDA Vernier resonator with a resonant mode number of 5;4;4 for (a) through port and (b) drop port.

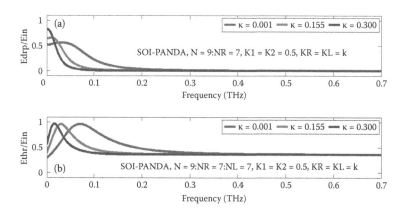

FIGURE 1.7 Optical transfer function of a PANDA Vernier resonator with a resonant mode number of 9;7;7 for (a) through port and (b) drop port.

of $k_t = 0.155$, the drop/input relative electric fields grow to 0.65 and for the frequency area larger than 0.26 THz, the value for drop port transfer function decreases to 0.1. Here the optical switching time is 6 ps. In this resonant mode, the optical switching time from 11 ps to 1.96 ps can be attained by altering the lateral coupling coefficients from 0.300 to 0.001, respectively. Consider the threshold coupling of $k_t = 0.155$ for through port of the PANDA Vernier resonator leads to a switching time of 4.8 ps while relative through/input port electric field after 0.26 THz gets the value of 0.36. For this mode, changing the lateral coupling coefficients from 0.001 to 0.300 gets a rise to the frequency switching of 1.88 ps and 5.9 ps, respectively.

Since ultrafast all-optical switches (AOSs) will be necessary in high-bit-rate communication systems [19] and all-optical switching devices based on ordinary optical fiber offer great promise for reducing the complexity of optical communication systems [38], attaining these fast frequency switching times make PANDA Vernier a proper component for fast switching keying applications [29–30].

All in all, the ultrafast optical switching time of 1.1 ps with insertion loss of 17 dB and interstitial suppression of 9.2 dB for the through port can be attained by adjusting the PANDA Vernier resonator on the resonant mode numbers of 5;4;4 with lateral and main coupling coefficients of 0.001 and 0.5, respectively. A resonant mode number of 9;7;7 is a convenient mode to achieve the fastest drop port switching time of 1.96 ps with insertion loss of 57 dB from the PANDA Vernier system with k1 = k2 = 0.5 and k = 0.001. Since the PANDA Vernier system does not need any pumping system, and its size is in the microrange, the PANDA Vernier resonator can be an operative configuration for optical signal processing in photonics integrated circuits.

1.4   CONCLUSION

A graphical approach is used to derive the optical transfer functions for the through port of the PANDA Vernier filter. The optical nonlinear phenomena such as bistability, the Ikeda instability, and the dynamic treatments of light traveling in silicon-on-insulator PANDA resonator for two resonant mode numbers are studied. The treatments of signals for the drop port of PANDA configurations with different mode numbers notably depend on symmetric coupling coefficients of $K_R = K_L$ for both modes. It is found that in addition to device parameters such as main and lateral coupling coefficients, mode resonant numbers also have a crucial effect on the through/input and drop/input relative powers for the SIO-PANDA filter. Simulation results obtained have shown that the shortest switching time of 1.1 ps for the through port with resonant mode of 5;4;4 and 1.96 ps for the drop port of the PANDA Vernier resonator with resonant mode of 9;7;7 can be achieved. The advantage of the proposed system is that no pumping component is required. Moreover, the dimension of the system is within the micrometer scale. Generally, the Vernier SOI-PANDA Vernier resonator is introduced as an effective configuration for optical signal processing, fast switching keying in photonics integrated circuits.

REFERENCES

1. C. L. Tang, A. Schremer, T. Fujita, Bistability in two-mode semiconductor lasers via gain saturation, *Appl. Phys. Lett,* 51, 1392–1394, 1987.
2. B. Li, M. I. Memon, G. Mezosi, Z. Wang, M. Sorel, and S. Yu, All-optical digital logic gates using bistable semiconductor ring lasers, *Journal of Optical Communications,* 30, 190–194, 2009.
3. M. Soljacic, M. Ibanescu, C. Luo, S. G. Johnson, S. Fan, Y. Fink, and J. D. Joannopoulos, All-optical switching using optical bistability in nonlinear photonic crystals, *SPIE,* 5000, 200–214, 2003.
4. S. Zhang, D. Owens, Y. Liu, M. Hill, D. Lenstra, A. Tzanakaki, G. D. Khoe, and H. Dorren, Multistate optical memory based on serially interconnected lasers, *Photonics Technology Letters, IEEE,* 17, 1962–1964, 2005.

5. A. Malacarne, J. Wang, Y. Zhang, A. D. Barman, G. Berrettini, L. Poti, and A. Bogoni, 20 ps transition time all-optical SOA-based flip-flop used for photonic 10 Gb/s switching operation without any bit loss, *Selected Topics in Quantum Electronics, IEEE Journal of,* 14, 808–815, 2008.

6. A. Bahrampour, S. Zakeri, S. Mohammad Ali Mirzaee, Z. Ghaderi, and F. Farman, All-optical set–reset flip-flop based on frequency bistability in semiconductor microring lasers, *Optics Communications,* 282, 2451–2456, 2009.

7. A. Bahrampour, M. Karimi, M. Qamsari, H. R. Nejad, and S. Keyvaninia, All-optical set–reset flip–flop based on the passive microring-resonator bistability, *Optics Communications,* 281, 5104–5113, 2008.

8. V. Van, T. Ibrahim, P. Absil, F. Johnson, R. Grover, and P. T. Ho, Optical signal processing using nonlinear semiconductor microring resonators, *Selected Topics in Quantum Electronics, IEEE Journal of,* 8, 705–713, 2002.

9. A. Costanzo-Caso Pablo, J. Yiye, S. Granieri, and A. Siahmakoun, Optical Bistability in a nonlinear-SOA-based fiber ring resonator, *Proc. of SPIE,* 7797, 9, 2010.

10. P. P. Yupapin and S. Suchat, Nonlinear penalties and benefits of light traveling in a fiber optic ring resonator, *Optik,* 120, 216–221, 2009.

11. F. Sanchez, Optical bistability in a 2×2 coupler fiber ring resonator: Parametric formulation, *Optics Communications,* 142, 211–214, 1997.

12. N. Dou and C. Li, Optical bistability in fiber ring resonator containing an EDFA, *Optics Communications,* 281, 2238–2242, 2008.

13. F. S. Felber and J. H. Marburger, Theory of nonresonant multistable optical devices, *Appl. Phys. Lett.,* 28, 731–733, 1976.

14. J. H. Marburger and F. S. Felber, Theory of a lossless nonlinear Fabry-Perot interferometer, *Phys. Rev. A,* 17, 335–342, 1978.

15. D. Miller, Refractive Fabry-Perot bistability with linear absorption: Theory of operation and cavity optimization, *Quantum Electronics, IEEE Journal of,* 17(3), 306–311, 1981.

16. K. Ikeda, H. Daido, and O. Akimoto, Optical turbulence: Chaotic behavior of transmitted light from a ring cavity, *Physical Review Letters,* 45, 709–712, 1980.

17. G. P. Agrawal, *Nonlinear Fiber Optics,* 2nd ed., Springer, Berlin, 1995.

18. H. Nakatsuka, S. Asaka, H. Itoh, K. Ikeda, and M. Matsuoka, Observation of bifurcation to chaos in an all-optical bistable system, *Phys. Rev. Lett.,* 50, 109–112, 1983.

19. J. Harbold, F. Ö. Ilday, F. Wise, J. Sanghera, V. Nguyen, L. Shaw, and I. Aggarwal, Highly nonlinear As-S-Se glasses for all-optical switching, *Optics Letters,* 27, 119–121, 2002.

20. S. J. Mason, Feedback theory—Some properties of signal flow graphs, *Proc. IRE,* 41, 1144–1156, 1953.

21. S. J. Mason, Feedback theory—Further properties of signal flow graphs, *Proc. IRE,* 44, 920–926, 1956.

22. B. Moslehi, J. W. Goodman, M. Tur, and H. J. Shaw, Fiber optic signal lattice processing, *Proc. IEEE,* 72, 909–929, 1984.

23. S. Mitatha, K. Dejhan, P. P. Yupapin, and N. Pornsuwancharoen, Chaotic signal generation and coding using a nonlinear micro ring resonator, *Optik-International Journal for Light and Electron Optics,* 121, 120–125, 2010.

24. P. P. Yupapin and W. Suwancharoen, Chaotic signal generation and cancellation using a micro ring resonator incorporating an optical add/drop multiplexer, *Optics Communications,* 280, 343–350, 2007.

25. P. P. Yupapin and N. Pornsuwancharoen, Proposed nonlinear microring resonator arrangement for stopping and storing light, *Photonics Technology Letters, IEEE,* 21, 404–406, 2009.

26. P. P. Yupapin, N. Pornsuwancharoen, and S. Chaiyasoonthorn, Attosecond pulse generation using the multistage nonlinear microring resonators, *Microwave and Optical Technology Letters,* 50, 3108–3111, 2008.

27. P. P. Yupapin, Coupler-loss and coupling-coefficient-dependent bistability and instability in a fiber ring resonator, *Optik-International Journal for Light and Electron Optics,* 119, 492–494, 2008.
28. N. Zou, W. Li, B. Huang, Z. Xu, S. Xu, and C. Yang, An optical continuous phase FSK modulation scheme with an arbitrary modulation index over long-haul transmission fiber link, *Optics Communications,* 285, 2591–2595, 2012.
29. A. Maurente, F. H. R. França, K. Miki, and J. R. Howell, Application of approximations for joint cumulative k-distributions for mixtures to FSK radiation heat transfer in multi-component high temperature non-LTE plasmas, *Journal of Quantitative Spectroscopy and Radiative Transfer,* 113, 1521–1535, 2012.
30. M. Bahadoran, A. Afroozeh, J. Ali, and P. P. Yupapin, Slow light generation using microring resonators for optical buffer application, *Optical Engineering,* 51, 044601-1–044601-8, 2012.
31. P. Saeung and P. P. Yupapin, Vernier effect of multiple-ring resonator filters modeling by a graphical approach, *Optical Engineering,* 46, 075005–075005-6, 2007.
32. S. Dey and S. Mandal, Modeling and analysis of quadruple optical ring resonator performance as optical filter using Vernier principle, *Optics Communications,* 285(4) 439–446, 2012.
33. P. P. Yupapin, P. Saeung, and C. Li, Characteristics of complementary ring-resonator add/drop filters modeling by using graphical approach, *Optics Communications,* 272, 81–86, 2007.
34. P. Saeung and P. P. Yupapin, Generalized analysis of multiple ring resonator filters: Modeling by using graphical approach, *Optik-International Journal for Light and Electron Optics,* 119, 465–472, 2008.
35. S. Mandal, K. Dasgupta, T. Basak, and S. Ghosh, A generalized approach for modeling and analysis of ring-resonator performance as optical filter, *Optics Communications,* 264, 97–104, 2006.
36. P. P. Yupapin and B. Vanishkorn, Mathematical simulation of light pulse propagating within a microring resonator system and applications, *Applied Mathematical Modelling,* 35, 1729–1738, 2011.
37. R. Boeck, N. A. Jaeger, N. Rouger, and L. Chrostowski, Series-coupled silicon race-track resonators and the Vernier effect: Theory and measurement, *Optics Express,* 18, 25151–25157, 2010.
38. J. E. Heebner and R. W. Boyd, Enhanced all-optical switching by use of a nonlinear fiber ring resonator, *Optics Letters,* 24, 847–849, 1999.

2 Optical Bistability

2.1 INTRODUCTION

Over the past few decades, optical ring resonators have attracted a considerable amount of attention as versatile elements for a wide range of applications ranging from telecommunication and sensing to basic scientific research [1,2]. The silicon-on-insulator (SOI) material system is a mature fabrication technique and provides the ability to implement microring resonators. In addition, the high confinement of light and the large third-order nonlinearity of silicon at wavelengths in the near infrared, typically around 1.4–1.6 μm, can be utilized for a number of applications by microring resonator devices [3,4]. The optical bistability observed in silicon-based microring resonators have been theoretically investigated and numerically simulated through various sophisticated techniques [5–9]. However, only recently unifying analytical theories regarding nonlinear effects of optical bistability in microring resonators in terms of a slowly varying envelope approach have been reported [10,11]. Previously, we analytically described the nonlinear response of microring resonators consisting of an add-drop filter with feedback and showed that Fano lineshapes can be adjusted by the resonator parameters [12]. In this chapter, we extend our work to examine the optical bistabililty in such microring resonators and test the analytical results with those predicted by the OptiFDTD software package.

In the sections that follow, a brief review of our analytical theories to understand the underlying physical mechanisms of optical nonlinearities in SOI microring resonators is presented. The analytical expressions for continuous-wave (CW) light propagating inside SOI microring resonators are tested with finite-difference time-domain (FDTD) simulations. We subsequently show that by varying the optical feedback length, the devices exhibit bistability and discuss the dependence of the bistability on linear and nonlinear parameters, such as the coupling coefficients, effective free-carrier lifetime, free-carrier absorption, and thermo-optic effect (TOE). Finally, the mode suppression by Vernier effect is demonstrated.

2.2 THEORETICAL MODEL

The geometry of the resonator structure investigated with the notation used is shown in Figure 2.1. The device consists of a single resonator channel dropping filter linked to a loop as a feedback structure. Four couplers are used in the device and are numbered consecutively in the counterclockwise direction. The perfect absorber is employed at the output of the feedback loop to protect the interference of light (I_{14}) to the system. We consider a CW signal at an angular frequency ω propagating inside a straight waveguide coupled vertically to a silicon ring of radius R. The evolution of the electric field associated with this optical wave is given by [10,11]

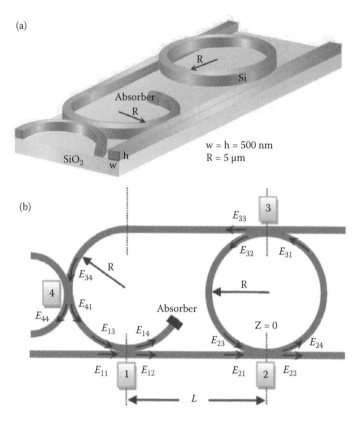

FIGURE 2.1 (a) Schematic diagram of the designed device and (b) details of the notation used.

$$\frac{1}{A}\frac{dA}{dz} = -\frac{\alpha}{2} - \left(\frac{\beta}{2} - i\gamma\right)|A|^2 - \left(\frac{\xi_r}{2} + i\xi_i\right)|A|^4 \qquad (2.1)$$

where $A(z)$ is the slowly varying envelope of the electric field related to the electric field as $E(z) = \varpi\, A(z)\, \exp(i\beta_0 z)$; $\varpi = (\mu_0/\varepsilon_0)^{1/4}(2n_0)^{1/2}$; μ_0 and ε_0 are, respectively, the permeability and permittivity of vacuum; n_0 is the linear refractive index, $\beta_0 = n_0 k$ is the propagation constant with $k = \omega/c$; and c is the speed of light in vacuum. The parameters in (2.1) are as follows: α, β, and $\gamma = k n_2$ govern linear losses, two-photon absorption (TPA), and the Kerr effect, respectively, n_2 being the nonlinear Kerr parameter.

The free-carrier coefficients ξ_r and ξ_i are defined in the SI system as [10,11]

$$\xi_r = \left(1.45 \times 10^{-21}\right)\left(\frac{\lambda}{1.55 \times 10^{-6}}\right)^2 \frac{\tau_{eff}\beta\lambda}{2hc} \qquad (2.2)$$

$$\xi_i = \left(5.3\times10^{-27}\right)\left(\frac{\lambda}{1.55\times10^{-6}}\right)^2 \frac{\tau_{eff}\beta}{2hc} - \frac{k\beta\kappa\vartheta}{C\rho} \tag{2.3}$$

where τ_{eff} is the effective free-carrier lifetime and h is Planck's constant. The second term of (2.3) accounts for the TOE, where κ, ϑ, C, and ρ are the thermo-optic coefficient, thermal dissipation time, thermal capacity, and density of silicon, respectively. The second terms of (2.1) can be discarded since TPA is much smaller than the free-carrier absorption (FCA) in the case of the CW signal [13]. The solution of (2.1), with $A(z) = \sqrt{I(z)}\exp[i\phi(z)]$, along both the straights and ring waveguides in Figure 2.1, are as follows [10,11]:

$$I(z) = \frac{I_0\exp(-\alpha z)}{\sqrt{1+I_0^2(\xi_r/\alpha)\left[1-\exp(-2\alpha z)\right]}} \tag{2.4}$$

$$\phi(z) = \phi_0 + \gamma I_0 L_{eff}(z) - \frac{\xi_i}{\xi_r}\left(\ln\frac{I_0}{I(z)} - \alpha z\right) \tag{2.5}$$

$$L_{eff}(z) = \frac{\tan^{-1}\left[I_0\sqrt{\xi_r/\alpha}\right] - \tan^{-1}\left[I(z)\sqrt{\xi_r/\alpha}\right]}{I_0\sqrt{\alpha\xi_r}} \tag{2.6}$$

where I_0 and ϕ_0 are the values of intensity and phase at $z = 0$, and L_{eff} is the generalized effective length. The second coupling between optical waves propagating in the ring and straight waveguide is assumed to group together at point $z = 0$. The electric field relation of each coupling region can be expressed as

$$E_{22} = r_2E_{21} + it_2E_{23} \tag{2.7}$$

$$E_{24} = it_2E_{21} + r_2E_{23} \tag{2.8}$$

$$E_{33} = it_3E_{31} \tag{2.9}$$

$$E_{44} = it_4E_{34} \tag{2.10}$$

$$E_{41} = r_4E_{34} \tag{2.11}$$

$$E_{14} = r_1E_{13} + it_1E_{11} \tag{2.12}$$

$$E_{12} = r_1E_{11} + it_1E_{13} \tag{2.13}$$

where $t_i = \sqrt{1-r_i^2}$ and r^2 is the fraction of power remaining in the straight waveguide after the coupler. Eliminating E_{21} from (2.7) and (2.8), we can express the intensity $I_{22} = |E_{22}|^2/\varpi^2$ in terms of $I_0 = |E_{24}|^2/\varpi^2$ as [10,11]

$$I_{22} = \frac{r_2^2 I_0 + r_3^2 I(2\pi R) - 2r_2 r_3 \sqrt{I_0 I(2\pi R)} \cos \Delta\phi_1}{1 - r_2^2} \qquad (2.14)$$

where $\Delta\phi_1 = 2\pi\beta_0 R + \phi(2\pi R) - \phi_0$ is the phase shift acquired during one round trip within the channel dropping microring resonator. A similar relation for the intensities I_{33}, I_{44}, and I_{41} can be obtained using (2.9) through (2.11)

$$I_{33} = (1 - r_3^2)I(\pi R) \qquad (2.15)$$

$$I_{44} = (1 - r_4^2)I_{34} \qquad (2.16)$$

$$I_{41} = r_4^2 I_{34} \qquad (2.17)$$

Let L denote the straight sections of the racetrack-shaped ring resonator. Then the intensity of the electric field along both the straight and the half-circle ring waveguides are given by [10,11]

$$I_{12} = \frac{I_{21} e^{\alpha L}}{\sqrt{1 + I_{21}^2 (\xi_r / \alpha) \left[1 - e^{2\alpha L} \right]}} \qquad (2.18)$$

$$I_{34} = \frac{I_{33} e^{-\alpha(L + \pi R/2)}}{\sqrt{1 + I_{33}^2 (\xi_r / \alpha) \left[1 - e^{-2\alpha(L + \pi R/2)} \right]}} \qquad (2.19)$$

$$I_{13} = \frac{I_{41} e^{-\alpha \pi R/2}}{\sqrt{1 + I_{41}^2 (\xi_r / \alpha) \left[1 - e^{-\alpha \pi R} \right]}} \qquad (2.20)$$

Eliminating E_{11} from (2.12) and (2.13), we can express the intensity $I_{14} = |E_{14}|^2 / \varpi^2$ as [10,11]

$$I_{14} = \frac{I_{13} + (1 - r_1^2)I_{12} + 2\sqrt{(1 - r_1^2)I_{12}I_{13}} \sin \Delta\phi_2}{r_1^2} \qquad (2.21)$$

where $\Delta\phi_2 = 2\pi\beta_0 R + 2\beta_0 L + \phi(2\pi R + L) - \phi(-L)$ is the phase shift acquired during one round trip within the racetrack-shaped ring resonator. From the energy conservation law, we obtain

$$I_{21} = I_{22} + I_0 - r_3^2 I(2\pi R) \qquad (2.22)$$

$$I_{11} = I_{14} + I_{12} - I_{13} \qquad (2.23)$$

In the presence of TPA, the exact solutions of (2.1) in the implicit form are given by [10,11,14]

$$2\alpha z + \beta I_0 L_{eff}(z) = \ln \frac{\alpha I^{-2}(z) + \beta I^{-1}(z) + \xi_r}{\alpha I_0^{-2} + \beta I_0^{-1} + \xi_r} \tag{2.24}$$

$$\phi(z) = \phi_0 + \gamma I_0 L_{eff}(z) - \frac{\xi_i}{\xi_r}\left[\ln \frac{I_0}{I(z)} - \alpha z - \beta I_0 L_{eff}(z)\right] \tag{2.25}$$

$$L_{eff}(z) = \frac{q}{\beta I_0} \ln\left[\frac{qK(z)+1}{qK(z)-1}\frac{qK(0)+1}{qK(0)-1}\right] \tag{2.26}$$

with $q = (1 - 4\alpha\xi_r/\beta^2)^{-1/2}$ and $K(z) = 1 + 2(\xi_r/\beta)I(z)$. Equations (2.24) through (2.26) give the evolution of the input–output relation in a parametric form, where the parameter I_0 lies in the intervals $[(0, \Im_1), (\Im_2, \Im_3), ..., (\Im_{n-1}, \Im_n)]$, where $\Im_j (j = 1, 2,..., n)$ are the roots of the preceding transcendental equations.

2.3   RESULTS AND DISCUSSION

First we look at the accuracy of analytical solutions when TPA and TOE are not taken into account. We performed FDTD simulations for the SOI microring devices of radius 5 μm, width 0.5 μm, refractive indices 3.484 for the core and 1.0 for the cladding, the lateral/vertical separation between the ring and straight waveguide of 0.15 μm, operating at wavelength 1.55 μm. The analytical calculation was determined by using parameters as follows: $R = 5.0$ μm; $L = 10.0$ μm; $n_0 = 3.484$, $\alpha = 1$ dB/cm, $\beta = 0.5$ cm/GW, and $n_2 = 6 \times 10^{-5}$ cm²/GW [10,11]. In Figure 2.2, we compare the throughput transmission intensity, plotted versus wavelength, in the absence of TPA and TOE (solid curve) with the nonlinear FDTD simulations (dashed curve). The correlation coefficient between the two in Figure 2.2a is 0.39. It is shown that the approximate analytical solutions are in agreement with the numerical FDTD data. A very close agreement between analytical solutions, when incorporating TOE and FDTD simulations, is shown in Figure 2.2b; in this case, the correlation coefficient between the two becomes 0.94. It should be noted that inclusion of the TOE results in red shifting the resonance peaks of the proposed device while the transmission spectra remain unchanged. As is evident from Figure 2.2, our theoretical model is capable of predicting the nonlinear phenomenon of the designed microring resonators.

Due to the nature of derived solutions (i.e., Equations 2.14 and 2.21), the optical bistability induced by nonlinear parameters can be anticipated. We initially investigated this nonlinear phenomenon by varying the length of feedback waveguides. Figure 2.3 reveals that the change of the bus length quantitatively affects the bistable behavior. The dotted curve in Figure 2.3 shows that bistability is absent when the length of straight bus $L = 10$ μm for input intensity in the range 0–5.0 GW/cm². The increase of the feedback length reduces the threshold intensity at which the

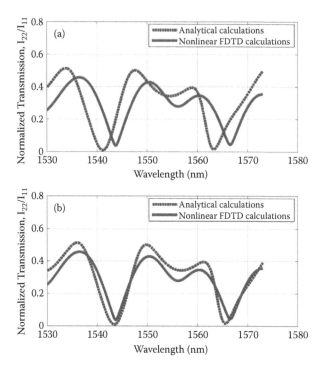

FIGURE 2.2 Transmission spectra for SOI microring resonators with feedback. Solid curves are determined from the analytic theory in (a) the absence and (b) the presence of TOE. Dashed curves are the numerical results using the FDTD method. The parameters used are $r_1 = 0.72$, $r_2 = 0.6$, $r_3 = 0.7$, $r_4 = 0.36$, $\tau_{\text{eff}} = 1$ ns, and $I_0 = 0.3$ GW/cm². For other parameters, see the text.

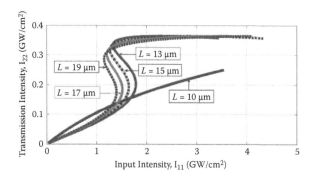

FIGURE 2.3 Bistable curves for five different bus lengths. The parameters are $R = 5$ μm, $L = 10$ μm, $r_1 = 0.6$, $r_2 = 0.3$, $r_3 = 0.3$, $r_4 = 0.6$, $\tau_{\text{eff}} = 1$ ns, and $\lambda = 1550$ nm. For other parameters, refer to the text.

bistable switching occurs and shrinks the width of the hysteresis loop. The reduction of the threshold for a longer feedback waveguide should result from the increased nonlinear interaction length. An arc waveguide with coupler 4 of Figure 2.1 permits the coupling of energy into and out of the rings. The coupling strength between the ring and arc waveguide can be manipulated by adjusting the gap during fabrication. Figure 2.4 plots the transmission intensity at the throughput as a function of input intensity for five different coupling parameters. As expected, the increase of coupling strength increases the light circulation inside the resonator and creates a higher nonlinear refractive index change, which results in the enlargement of hysteresis width. We note that between $r_4 = 0.3$ and $r_4 = 0.4$ the optical system enters the bistability regime.

Free carriers (FCs) generated by TPA cause optical losses (FCA) and nonlinear phase shifts (FCD), which in turn lower the peak intensity inside the waveguide and results in a shifting of the resonance peaks. The effect of FCs generated through TPA on optical bistability can be inferred from the effective FC's lifetime. To illustrate this impact, we plot transmission intensity for different effective FC lifetimes at 1500 nm (see Figure 2.5). As expected, increasing the lifetime of FCs decreases transmission intensity and reduces the switching ratio due to the strong nonlinear

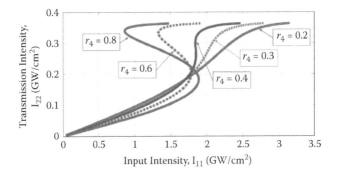

FIGURE 2.4 Bistable characteristics for different values of the coupling coefficients r_4. The parameters are shown in Figure 2.3 but the length of the bus waveguides is fixed to $L = 13$ μm.

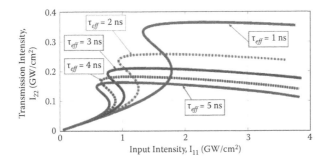

FIGURE 2.5 The effect of free carriers generated through TPA on bistable curves. The parameters are shown in Figure 2.4.

absorption. Furthermore, it reduces input intensities at which bistable switching occurs. Next, we examine the impact of the operating wavelength on the features of the bistability curves. As illustrated in Figure 2.6, the bistable behavior of the device depends drastically on the operating wavelength (λ). At the operating wavelength $\lambda = 1550$ nm, the I_{22} versus I_{11} graph exhibits a self-intersecting hysteresis loop. A decrease in wavelength from 1550 to 1544 nm not only increases the threshold switching intensities and multibistability, but collapses the loop-shaped hysteresis. We speculate that the loop-shaped hysteresis arises due to a mismatch between the internal and output resonances, which are consistent with those of other studies [15,16]. Finally, we focus on mode suppression of the designed device by Vernier effect for tunable filter applications. The Vernier concept for the designed ring resonators is realized through the appropriate selection of ring coupling coefficients $r_2 = 0$ and $r_4 = 0$ as illustrated in an inset of Figure 2.7. An input signal is filtered by a first ring having the free spectral range (FSR) of FSR1 = 11 ns (1.4 GHz). The output of the first ring is then filtered by a second ring having the FSR of FSR2 = 9 ns (1.2 GHz). The combined action of these two rings results in dramatically enhanced FSR up to 53 ns (FRS3 = 6.9 GHz).

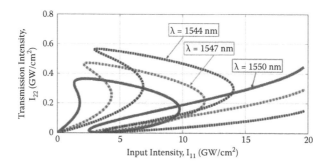

FIGURE 2.6 Bistable curves for three different operating wavelengths. The parameters are shown in Figure 2.4.

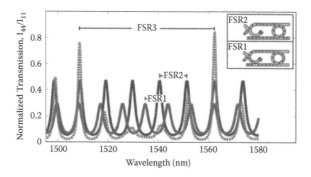

FIGURE 2.7 Mode suppression by the Vernier effect. Main parameters are shown in Figure 2.4.

2.4 CONCLUSION

In this work, we have presented a validation analysis of our previously published analytical theory for the optical effects in SOI microring resonators. The results predicted by our theory are tested with commercial FDTD simulations for CW input. A very good agreement is found between the theory and the simulation results. Subsequently, we investigated the phenomenon of optical bistability and hysteresis in detail. The bistable switching threshold and the hysteresis loop width are found to depend very sensitively on the coupling coefficients, length of the feedback loop, T_{eff}, and thermo-optic effect. One unanticipated finding is that the loop-shaped hysteresis occurs at the communication wavelength. We have also presented the possibility to selectively suppress modes of the transmission spectra of the designed device by utilizing the Vernier effect.

REFERENCES

1. J. Heebner, R. Grover, and T. Ibrahim, *Optical Microresonators: Theory, Fabrication and Applications*, 1st ed., London, UK: Springer-Verlag, 2008.
2. O. Schwelb, Transmission, group delay, and dispersion in single-ring optical resonators and add/drop filters—A tutorial overview, *J. Lightw. Technol.*, 22 (5) 1380–1394, 2004.
3. W. Bogaerts, P. D. Heyn, T. V. Vaerenbergh, K. D. Vos, S. K. Selvaraja, T. Claes, P. Dumon, P. Bienstman, D. V. Thourhout, and R. Baets, Silicon microring resonators, *Laser Photon.*, 6 (1) 47–73, 2012.
4. R. Dekker, N. Usechak, M. Forst, and A. Driessen, Ultrafast nonlinear all-optical processes in silicon-on-insulator waveguides, *J. Phys. D: Appl. Phys.*, 40 (14) R249–R271, 2007.
5. J. Petráček, A. Sterkhova, and J. Luksch, Numerical scheme for simulation of self-pulsing and chaos in coupled microring resonators, *Microw. Opt. Technol. Lett.*, 53 (10) 2238–2242, 2011.
6. V. Van, T. A. Ibrahim, P. P. Absil, F. G. Johnson, R. Grover, and P. T. Ho, Optical signal processing using nonlinear semiconductor microring resonators, *IEEE J. Sel. Top. Quantum Electron.*, 8 (3) 705–713, 2002.
7. S. Chen, L. Zhang, Y. Fei, and T. Cao, Bistability and self-pulsation phenomena in silicon microring resonators based on nonlinear optical effects, *Opt. Exp.*, 20 (7) 7454–7468, 2012.
8. Y. Chen and S. Blair, Nonlinearity enhancement in finite coupled-resonator slow-light waveguides, *Opt. Exp.*, 12 (15) 3353–3366, 2004.
9. Y. Dumeige and P. Féron, Dispersive tristability in microring resonators, *Phys. Rev. E.*, 72, 1–8, 2005.
10. I. D. Rukhlenko, M. Premaratne, and G. P. Agrawal, Analytical study of optical bistability in silicon-waveguide resonators, *Opt. Exp.*, 17 (24) 22124–22137, 2009.
11. I. D. Rukhlenko, M. Premaratne, and G. P. Agrawal, Analytical study of optical bistability in silicon ring resonators, *Opt. Lett.*, 35 (1) 55–57, 2010.
12. S. Pitakwongsaporn and S. Chiangga, Tunable asymmetric fano lineshapes in silicon-based microring resonators with feedback, *J. Nonlinear Opt. Phys. Mater.*, 20 (3) 357–366, 2011.
13. Q. Lin, O. J. Painter, and G. P. Agrawal, Nonlinear optical phenomena in silicon waveguides: Modeling and applications, *Opt. Exp.*, 15 (25) 16604–16644, 2007.
14. I. D. Rukhlenko and M. Premaratne, Spectral compression and group delay of optical pulses in silicon Raman amplifiers, *Opt. Lett.*, 35 (18) 3138–3140, 2010.

15. D. C. Hutchings, A. D. Lloyd, I. Janossy, and B. S. Wherrett, Theory of optical bistability in metal-mirrored Fabry–Perot cavities containing thermo-optic materials, *Opt. Commun.*, 61 (5) 345–350, 1987.

16. D. N. Maywar and G. P. Agrawal, Effect of chirped gratings on reflective optical bistability in DFB semiconductor laser amplifiers, *IEEE J. Quantum Electron.*, 34 (12) 2364–2370, 1998.

3 Fast, Slow, Stopping, and Storing Light

3.1 INTRODUCTION

The behavior of light in nano/micro optical devices has shown promising results which can be useful for much fundamental and applied research, especially in nanoelectronics. In particular, with the manipulation of fast and slow light, Pornsuwancharoen and Yupapin [1] have proposed the use of a simple device called a *microring resonator* to perform such behavior, in which the four different behaviors of light, that is, fast, slow, stopping, and storing of light are investigated using a ring resonator. Nowadays, stopping or cooling a light beam has become a promising technique for atom/molecule trapping investigations (using static or dynamic tweezers), especially after the announcement of the 2012 Nobel Prize in Physics for the whispering gallery modes [2,3]. There are two more kinds of devices that are used to trap light beams: the use of microcavity arrays performed by Yanik and Fan [4], and the nonlinear microring resonator by Yupapin and Pornsuwancharoen [5] for stopping light (laser beam). Ang and Ngo have also recently done experiments in slowing light in microresonators using a microring system [6]. This concept is a concrete backbone for many applications.

Electromagnetic waves can be confined strongly within the structure known as whispering gallery modes (WGMs), which has been widely studied in the form of sound, optical, and microwave wavelengths in various systems, including cylindrical and toroidal geometries, micrometer-sized liquid droplets, and glass and crystal spheres. In recent years, WGM resonators (WGMRs) have attracted scientists and researchers, and a large amount of research work has been done, from the simplest structures, namely cylindrical optical fibers [7] and microspheres [8], microdisks [9], microtoroids [10,11], photonic crystal cavities [12], microcapillaries [13] to the interesting structures, such as bottle [14] and bubble [15] microresonators. The important aspects of integrated circuit and cost-saving mass manufacturing processes have also concentrated much research on planar waveguide structures such as ring resonators and disk resonators devices, which can be microfabricated onto wafer substrates using conventional integrated-circuit deposition and etching techniques.

In recent years there has been activity aimed at developing the techniques that can lead to a significant modification of the group velocity of propagating a light pulse through a medium. Proposed applications for these procedures include the development of optical slow light, fast light, stop light, and the *storage* of light pulses (photonic storing) with perhaps implications for the field of quantum information. Most of all research proposes the use of the response of resonant media and the concept of electromagnetically induced transparency. A PANDA ring resonator was

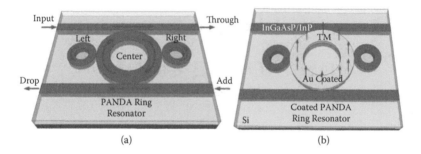

(a)   (b)

FIGURE 3.1 A conventional PANDA ring planar waveguide (named and designed by K. Uomwech, K. Sarapat, and P. P. Yupapin, Dynamic Modulated Gaussian Pulse Propagation within the Double PANDA Ring Resonator System, *Microw. & Opt. Techn. Lett.*, 52 (8) 1818–1821, 2010), and a gold coated PANDA ring resonator for TM polarized coupling device.

demonstrated by Yupapin and his colleagues, which is formed by using an all-optical device and consists of a traditional add/drop optical filter (linear device) modified with double side rings as shown in Figure 3.1. The PANDA ring resonators can be of arbitrary nonlinear device design, where a whispering gallery mode is an interesting aspect. In this system, the light field effectively circulates many times in each resonator; therefore, the group velocity of a propagating light pulse through such a structure is significantly reduced.

In addition, the phase shift is experienced by the light wave while interaction with each resonator depends sensitively on its detuning from the cavity resonance, and thus this structure produces large and controllable dispersion. Moreover, if the resonator is made up of a material that displays a nonlinear optical response, the nonlinear phase shift acquired by a light wave interacting with the resonator scales as the square finesse of the resonator, then such structure displays an enhanced nonlinear optical response. There are countless applications for the PANDA ring resonator, such as quantum coding, nano-antennas, biomedical, optical tweezer, stop light, and so forth, which will play an important role in a future optical device.

In this chapter, a new design of microring resonator device is proposed which can be used to generate four forms of light simultaneously on a chip, while the storing and harvesting of trapped atoms/molecules can also be available. The proposed device is made up of silica and InGaAsP/InP with a linear optical add-drop filter incorporating two nonlinear micro/nano rings on both sides of the center ring (add-drop filter). This particular configuration is known as a PANDA ring resonator [16] as shown in Figure 3.1. Light pulse, for instance, Gaussian, and bright and dark solitons are fed into the system through different ports such as an add port and through port. By using the practical device parameters, the simulation results are obtained using the Optiwave and MATLAB™ programs. Results obtained by both analytical and numerical methods show that many applications can be exploited. In simulation, the practical device parameters are used; therefore, such a device can be fabricated and implemented in the near future.

The proposed system can be used for many applications, especially when the device is coated with a metallic material [17]. The applications such as atom/molecule transportation, molecule/atom antenna, atomic/molecular automata, atom/molecule electronic devices, cells/atom distributed sensors, cells/atom radio, cell communications, micro-Faraday cage, molecular/atom storage and logistics, atom/molecule trapping (storage) or cooling, everlasting atom/molecule investigation, atomic/molecular storage for quantum gate or computer application, storm search and navigator sensors, magnetic/spin networks, magnetic net, microplasma source, 3-D flat panel device, and large cooling area (volume) can be constructed in the same way as a single device. In this work one of them is demonstrated. Under the stopping and storing condition, the system is considered a storage unit, where the atom or molecule can be trapped by the whispering gallery light beam at the center.

3.2   THEORETICAL BACKGROUND

Optical whispering gallery modes were first explained by Lord Rayleigh to excite mechanical whispering gallery modes within microresonators. Manipulation of sound waves inside a cavity involved the physics of resonance based on wave interference; these mechanical vibrations were excited via the interaction between photoelastic scattering of light by sound and electrostrictive forces by light on sound. It has applications in different fields such as the realization of microlasers, narrow filters, optical switching, ultrafine sensing, displacement measurements, high-resolution spectroscopy, Raman sources, and studies of nonlinear optical effects. Optical resonators play an important role in optical technologies, which can be fabricated by exploiting either total internal reflection (TIR) of light at the interface between a dielectric material and the surrounding air, or distributed Bragg reflection (DBR) from periodical structures such as multilayered structures or arrays of holes. The spectra of optical modes supported by microresonators are shape and size dependent. It is quite suitable to use and realize with a compact size, high mode quality factor (Q), and large free spectral range (FSR).

The sketch of a conceptual PANDA ring resonator is shown in Figure 3.1. The electromagnetic field in the WGM can be described by time-dependent Maxwell's equations. In source-free nonconducting media, the wave equation can be written as:

$$\nabla^2 \vec{E} - \frac{n^2}{c^2}\frac{\partial^2 \vec{E}}{\partial t^2} = 0 \tag{3.1}$$

where n is the refractive index of the media and c is the speed of light in free space. For time-harmonic fields such as $\vec{E}(\vec{r},t) = \vec{E}(\vec{r})e^{-i\omega t}$, Equation (3.1) is rewritten as:

$$\nabla^2 \vec{E} + \frac{n^2\omega^2}{c^2}\vec{E} = 0 \tag{3.2}$$

When the resonance ring is placed in the (x, y) plane as shown in Figure 3.1 and the electric field is polarized along the z-direction, the transverse electric (TE) field is given by $\vec{E}(\vec{r}) = E(x, y)\vec{z}$. In this case, Equation (3.2) becomes a scalar Helmholtz equation:

$$\nabla^2\vec{E} + n^2k_0^2\vec{E} = 0 \tag{3.3}$$

where k_0 is the free-space wave number.

For simplicity, the resonance ring is placed on the same plane as the waveguide. Thus, WGM can be treated in a two-dimensional model. For outside boundaries, the scattering boundary condition is applied with an initial amplitude of $E_0 = 0$. For laser excitation source, the scattering boundary condition is also used with input amplitude of $E_0 = 1$V/m. The scattering boundary condition is commonly used to specify a boundary which is transparent for a scattered wave and for an incoming plane wave. Equation (3.3) as $(\nabla^2 + k_0^2)E_z = 0$ in z-direction can be written in cylindrical coordinates as:

$$\left(\frac{\partial^2}{\partial r^2} + \frac{1}{r}\frac{\partial}{\partial r} + \frac{1}{r^2}\frac{\partial^2}{\partial\varphi^2} + k_0^2\right)E_z(r, \varphi) = 0 \tag{3.4}$$

where r is the radial distance, φ is the angle measured counterclockwise from the polar axis to the ray from the origin. By using the separation method, Equation (3.4) can be separated and split into two equations, (i) radial and (ii) azimuthal components. An integer m is introduced which is connected to the two equations and corresponded to the number of optical cycles. The azimuthal equation is given as

$$\left(\frac{\partial^2}{\partial\varphi^2} + m^2\right)E_z(\varphi) = 0 \tag{3.5}$$

and the solution of this equation is the complex exponentials, which is $E_z(\varphi) = e^{\pm im\varphi}$. The radial equation is the Bessel's function, which is given as

$$\left(\frac{\partial^2}{\partial\varphi^2} + \frac{1}{r}\frac{\partial}{\partial r} + k_0^2 - \frac{m^2}{r^2}\right)E_z(r) = 0 \tag{3.6}$$

where the solutions of Bessel's equations are the Bessel's function of the first J_m and the second Y_m kinds. As the second function is singular at the origin, only the first kind of function is retained inside the ring, whereas outside the ring both functions are well behaved and must be retained. The Hankel functions are a linear superposition of two Bessel function solutions corresponding to outward, $H_m^1 = J_m + iY_m$ and inward, $H_m^2 = J_m - iY_m$ propagating cylindrical waves. The analysis of the wave arriving at the resonator from the radial horizon is not considered here, and thus only the Hankel function of the first kind, H_m^1, is retained. Thus, the appropriate solution for the radial field dependence for both interior (r < R) and exterior (r > R) to the dielectric ring are given as

$$E_z(r < R) = A_m J_m(\bar{k}_1 r) \tag{3.7}$$

$$E_z(r > R) = B_m H_m^1(\bar{k}_2 r) \tag{3.8}$$

Here, $\bar{k}_1$ and $\bar{k}_2$ are complex propagation constants.

The complete axial electric field for the interior and exterior to the ring is constructed from the azimuthal and radial solutions including the boundary condition at the interface (r = R) which forces the tangential electric field to be continuous:

$$E_z(r,\varphi) = A_m J_m(\bar{k}_1 r) e^{i(\pm m\varphi - \omega t)} \tag{3.9}$$

$$E_z(r,\varphi) = A_m \frac{J_m(\bar{k}_1 R)}{H_m^1(\bar{k}_2 R)} H_m^1(\bar{k}_1 r) e^{i(\pm m\varphi - \omega t)} \tag{3.10}$$

In this proposal, we introduce the PANDA ring resonator as shown in Figure 3.2. Theoretical review for this concept is presented as the following: the relative phase of the two output light signals after coupling into the optical coupler is $\pi/2$. The signals coupled into the drop port and the through port obtained a phase difference of π with respect to the input port signal. The input and control fields at the input port and control port are described as follows [7]:

$$E_{Th} = x_1 y_1 E_{in} + \left(jx_1 x_2 y_2 \sqrt{\kappa_1} E_4 E_l E_1 - x_1 x_2 \sqrt{\kappa_1 \kappa_2} E_R E_{ad} \right) e^{-\frac{\alpha}{2}\frac{L}{2} - jk_n \frac{L}{2}} \tag{3.11}$$

$$E_{Dr} = x_2 y_2 E_{ad} + jx_2 \sqrt{\kappa_2} E_r E_1 e^{-\frac{\alpha}{2}\frac{L}{2} - jk_n \frac{L}{2}} \tag{3.12}$$

Where $x_1 = \sqrt{1-\kappa_1}$, $y_1 = \sqrt{1-\gamma_1}$, $x_2 = \sqrt{1-\kappa_2}$, $y_2 = \sqrt{1-\gamma_2}$, κ_1 and κ_2 is the intensity coupling coefficient, γ_1 and γ_l are the fractional coupler intensity loss, α is the attenuation coefficient, $\kappa_n = 2\pi/\lambda$ is the wave propagation number, λ is the input wavelength light field and $L = 2\pi R_{ad}$, where R_{ad} is the radius of the add-drop device. The circulated light fields, E_l and E_r, are the light fields circulated components of the nanoring radii, R_l and R_r, which are coupled to the left and right sides of the add-drop optical multiplexing system, respectively. The light field is transmitted and circulated in the right nanoring.

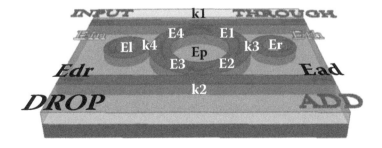

FIGURE 3.2 A schematic of a PANDA ring resonator.

The characteristic lengths of surface plasmon polaritons (SPPs), namely, the SPPs wavelength (λ_{spp}), propagation length (L_{spp}), and confinement to the surface (L_z) can be derived from the SPPs complex wavenumber, which is given as

$$\kappa_{spp} = \frac{\omega}{c}\sqrt{\frac{\varepsilon_c\varepsilon_d}{\varepsilon_c + \varepsilon_d}}, \tag{3.13}$$

where ω is the angular frequency, c the speed of light in vacuum, and ω_c and ω_d are the relative permittivity of the conductor and the dielectric, respectively. For simplicity, the vacuum is considered as the dielectric with $\varepsilon_d = 1$. The relative permittivity of the conductor is a complex quantity, $\varepsilon_c = \Re(\varepsilon_c) + i\Im(\varepsilon_c)$, which can be approximated by a Drude model for free-charge carriers as [18]

$$\varepsilon_c = \varepsilon_\infty\left(1 - \frac{\omega_P^2}{\omega^2 + i\gamma\omega}\right), \tag{3.14}$$

where ε_d is the high-frequency permittivity, γ is the average collision rate of the charge carriers, and $\omega_P = \sqrt{Ne^2/\varepsilon_\infty\varepsilon_0 m^*}$ is the plasma frequency. The waveguide charge carrier concentration is given by N, while e is the fundamental charge, ε_0 the vacuum permittivity, and m^* is the charge effective mass. Therefore, the use of SPPs mode coupling has to be confined well for transverse magnetic (TM) mode in the PANDA ring resonator.

3.3   SIMULATION RESULTS

The whispering gallery mode result is obtained by using the Optiwave program as shown in Figure 3.3. The ring material is InGaAsP/InP, where the device parameters are given in the figure caption. By using the MATLAB program, the whispering gallery modes of the four states of light, that is, fast, slow, stopping, and storing can be generated and controlled simultaneously on-chip as shown in Figure 3.4. The storing

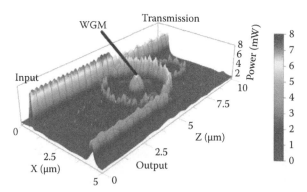

FIGURE 3.3   Results of whispering gallery mode of light within a PANDA ring waveguide InGaAsP/InP, $R_1 = R_2 = 0.775\,\mu m$, $R_{ad} = 1.565$, $A_{eff} = 0.3\,\mu m^2$, $n_{eff} = 3.14$, $n_2 = 1.3 \times 10^{-13}\,cm^2/W$, $\kappa_1 = \kappa_2 = \kappa_3 = \kappa_4 = 0.5$, $\gamma = 0.01$, $\lambda_0 = 1550$ nm.

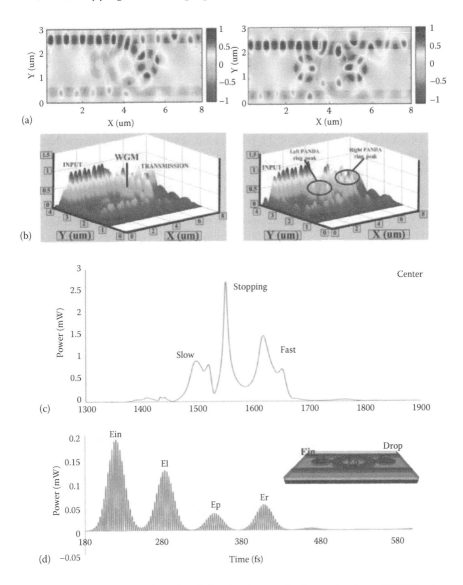

FIGURE 3.4 Stopping and storing light simultaneously detected using a PANDA ring, where (a) center ring, (b) side rings, (c) center peak and side peaks, and (d) fast (2nd peak) and slow (4th peak) light detected at drop port with time interval of 150 fs.

stage can be seen easily, while the stopping condition can be observed by satisfying the following conditions: (i) the center signal is lost in time between fast and slow signals or (ii) there is no movement among trapped particles or molecules, that is, the exchange of angular momentum introduces the conservation of angular momentum, where the combination of scattering and gradient forces is balanced under the adiabatic process.

Stopping light in terms of signal condition can be easily performed using the whispering gallery mode concept, where the fast and slow light can be used as the upper and lower time frames or upper side and lower side peak signals for the storing light at the center as shown in Figure 3.4, where in this case the movement (modulated signals) longer than 150 fs, that is, ms, ns, ps is observed (stopped). The input pulse is a Gaussian pulse with pulse width of 100 fs, the fast and slow time interval is known; however, the whispering gallery modes can be seen only under the resonant condition.

In this chapter, the full-wave finite-difference time-domain (FDTD) method is employed to solve Maxwell's equations. The FDTD modeling gives a useful design role, which is the combination of the propagation, scattering, diffraction, reflections, and polarization effects. It also handles well the material anisotropy, dispersion, and nonlinearities without any presumption of field behavior as the slowly varying amplitude approximation [19]. When a two-dimensional (2-D) problem is included, the photonic device is laid out in the x–z plane, the propagation is along the y-axis, and the z-direction is assumed to be infinite and there is no variation in the z-direction. This assumption removes all the $\partial/\partial z$ derivatives from Maxwell's equations and splits them into two (TE and TM) independent sets of equations. The location of the fields in the 2-D computational domain is shown in Figure 3.5. The space steps in the x- and y-directions are ~x and ~y, respectively. When a Gaussian optical pulse with a narrow bandwidth is applied to excite only one mode, the mode field distribution by the FDTD simulation can be obtained. Figure 3.6 depicts the distribution of E and H, for TE and TM mode, respectively. Because the field distribution is symmetrical about the z = 0 plane, we only present the field distribution at z > 0, where we find that the field distribution oscillates in the cladding layer for TE mode, but is still confined well for TM mode. So, the mode coupling between TE WGM and the vertical propagating mode in the PANDA induces the radiation loss and a low Q-factor for TE WGMs [20]. Although the mode wavelengths of TM WGMs are usually larger than the cutoff wavelengths of the radiated HE and EH modes with the same azimuthal and radial mode numbers in the microcylinder, so the corresponding mode-coupling radiation loss is absent for the TM WGMs. In our FDTD simulation, the perfectly matched layer (PML) absorbing boundary conditions applied by the Berenger [21] and Yee scheme [22] absorb

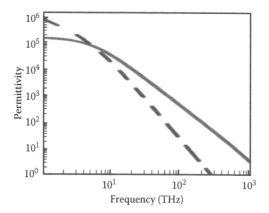

FIGURE 3.5 Complex permittivity ε_c of gold as a function of frequency, where the solid lines represent $-\Re(\varepsilon_c)$, while the dashed line is $\Im(\varepsilon_c)$.

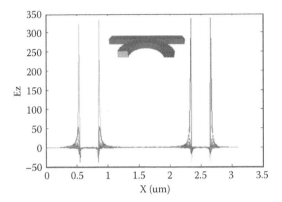

FIGURE 3.6 Surface energy confinement in a PANDA ring resonator.

the electromagnetic wave without any reflection at the computational boundary. A 100 fs Gaussian pulse modulated by a 200 THz carrier is exited. The vertical waveguide thickness and material composition is accounted for by computing the effective refractive index n_{eff} for the fundamental mode at $\lambda = 1.55$ μm and then using n_{eff} as the bulk material index of the core in the two-dimensional simulations. In the vertical direction, each waveguide structure is 0.45 μm thick, vertical core thicknesses are of 0.3 μm to 0.5 μm, and n_{eff} is between 3.2 to 3.4, in which the parameters are obtained by using the related practical material parameters of InGaAsP/InP. Here, the waveguide core n = 3.14 is bordered on each side by air n = 1. The parameters for add-drop optical multiplexer and both nanorings on the left- and right-hand sides of the PANDA ring are set at $R_1 = Rr = 0.775$ μm, the radius of the center ring is $R_{ad} = 1.56$ μm. The coupling coefficient ratios are $\kappa_1 = \kappa_4 = 0.5$, $\kappa_2 = \kappa_3 = 0.5$, effective core area of the waveguides is $A_{eff} = 0.25$ μm², and waveguide loss coefficient is $\alpha = 0.1$ dB/mm.

When gold material is coupled as a function of frequency by coating on the waveguide as shown in Figure 3.1, the free-carrier concentration of gold is $N \approx 6 \times 10^{22}$ cm⁻³ and the plasma frequency is $\omega_p/2\pi \approx 2 \times 10^3$ THz [23–24]. At high frequencies, in the visible regime, the real and imaginary components of the permittivity of gold has a small value. As the frequency decreases, the permittivity increases due to the ω^{-2} dependence in Equation 3.14. At 200 THz frequency, the permittivity has large absolute values as shown in Figure 3.5. The use of a light-trapping probe for atom/molecule trapping and transportation (dynamically trapping) can be formed with a wide range of applications. In this case, the modulated signal is required to switch off the whispering gallery mode power via the add port, where atoms/molecules at the device center can be trapped and transported along the waveguide by the surface plasmon tweezers as shown in Figure 3.7. The dynamic tweezers are generated by a PANDA ring, where in practice, particle angular momentum can be introduced by a metal coating material on the waveguide surface or combining the external modulation via the add port, which can be used to trap and transport the atom/molecule to the required destination when the gradient force is greater than the scattering force along the waveguide. For example, the use of such a concept for a new type of solar cell is shown in Figure 3.8, where free

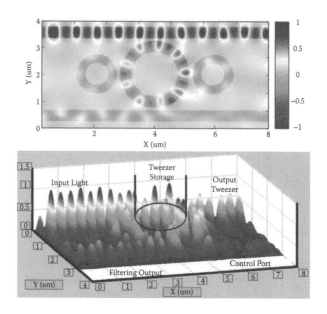

FIGURE 3.7 Dynamic tweezers are generated by a PANDA ring and transmitted via a through port for atoms/molecules harvesting and transportation.

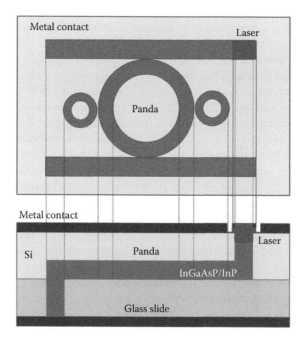

FIGURE 3.8 New type of solar cells using a PANDA and embedded particle accelerator (optical trapping and transportation). (From I. Srithanachai, S. Ueamanapong, S. Niemcharoen, and P. P. Yupapin, Novel Design of Solar Cell Efficiency Improvement Using an Embedded Electron Accelerator On-Chip, *Optics Express*, 20 (12) 12640–12648, 2012.)

electrons from the depletion region can be trapped and transported (injected) to the metal contact faster than the conventional device; the solar cell's efficiency can be increased by five times to the conventional one.

3.4 CONCLUSION

Forms of light in a PANDA ring resonator with or without coated material have been manipulated, and it has been observed that the two nonlinear side rings have shown interesting results and aspects. The input light can be in the form of soliton or Gaussian pulses. The use of photon or matter wave as input is also possible; an especially interesting aspect can also be formed by using the trapped electron, in which the matter wave concept is established within the PANDA ring waveguide. In this work, we found that four behaviors of light, for instance, fast, slow, stopping, and storing can be manipulated and seen simultaneously by using the PANDA ring planar waveguide, which can be fabricated and tested on-chip. The expected output light can be in the form of surface plasmon, potential wells, leaky modes, whispering gallery modes, matter wave, and photons (particles). The use of the nonlinear Schrödinger's equation will be our continuing investigation, where in this case the propagation of light is treated as a particle (photon) within the PANDA ring, in which the tunneling effects of particles can be performed and investigated. One more interesting aspect can also be formed by using the modulated signals input via the add port, which will be continued in our work.

REFERENCES

1. N. Pornsuwancharoen and P. P. Yupapin, Generalized fast, slow, stop and store light optically within a nanoring resonator, *Microw. & Opt. Techn. Lett.*, 51 (4) 899–902, 2009.
2. D. J. Wineland, J. J. Bollinger, W. M. Itano, and J. D. Prestage, Angular momentum of trapped atomic particles, *JOSA B*, 2 (11) 1721–1730, 1985.
3. J. C. Knight, N. Dubreuil, V. Sundoghdar, J. Hare, V. Lefevre-Seguin, J. M. Raimond, and S. Haroche, Characterizing whispering-gallery modes in microspheres by direct observation of the optical standing-wave pattern in the near field, *Opt. Lett.*, 21 (10) 698–700, 1996.
4. M. F. Yanik and S. Fan, Stopping and storing light coherently, *Phys. Rev. Lett.*, 92, 083901–3, 2004.
5. P. P. Yupapin and N. Pornsuwancharoen, Proposed nonlinear microring resonator arrangement for stopping and storing light, *IEEE Photon. Techn. Lett.*, 21 (6) 404–406, 2009.
6. T. Y. L. Ang and N. Q. Ngo, Tunable flat-band slow light via contra-propagating cavity modes in twin coupled microresonators, *JOSA B*, 29 (5) 924–933, 2012.
7. T. A. Birks, J. C. Knight, and T. E. Dimmick, High-resolution measurement of the fiber diameter variations using whispering gallery modes and no optical alignment, *IEEE Phot. Techn. Lett.*, 12, 182–183, 2000.
8. A. Chiasera, Y. Dumeige, P. Feron, M. Ferrari, Y. Jestin, G. N. Conti, S. Pelli, S. Soria, and G. C. Righini, Spherical whispering-gallery-mode microresonators, *Laser Photon. Rev.* 4, 457–482, 2010.
9. Y. H. Chen, Y. K. Wu, and L. J. Guo, Photonic crystal microdisk lasers, *Appl. Phys. Lett.* 98, 131109, 2011.
10. C. Zheng, X. Jiang, S. Hua, L. Chang, G. Li, H. Fan, and M. Xiao, Controllable optical analog to electromagnetically induced transparency in coupled high-Q microtoroid cavities, *Opt. Express*, 20, 18319–18325, 2012.

11. D. K. Armani, T. J. Kippenberg, S. M. Spillane, and K. J. Vahala, Ultra-high-Q toroid microcavity on a chip, *Nature*, 421, 925–928, 2003.

12. Y. Gonga and J. Vučković, Photonic crystal cavities in silicon dioxide, *Appl. Phys. Lett.*, 96, 031107-3, 2010.

13. K. Scholten, X. Fan, and E. T. Zellers, Microfabricated optofluidic ring resonator structures, *Appl. Phys. Lett.*, 99, 141108-3, 2011.

14. G. S. Murugan, J. S. Wilkinson, and M. N. Zervas, Selective excitation of whispering gallery modes in a novel bottle microresonator, *Opt. Express*, 17, 11916–11925, 2009.

15. H. Li, Y. Guo, Y. Sun, K. Reddy, and X. Fan, Analysis of single nanoparticle detection by using 3-dimensionally confined optofluidic ring resonators, *Opt Express*, 18, 25081–25088, 2010.

16. K. Uomwech, K. Sarapat, and P. P. Yupapin, Dynamic modulated Gaussian pulse propagation within the double PANDA ring resonator system, *Microw. & Opt. Techn. Lett.*, 52 (8) 1818–1821, 2010.

17. N. Thammawongsa, N. Moonfangklang, S. Mitatha, and P. P. Yupapin, Novel nano-antenna system design using photonics spin in a panda ring resonator *PIER L*, 31, 75–87, 2012.

18. M. van Exter and D. Grischkowsky, Optical and electronic properties of doped silicon from 0.1 to 2 THz, *Appl. Phys. Lett.*, 56, 1694–1696, 1990.

19. E. Waks and V. Jelena, Coupled mode theory for photonic crystal cavity-waveguide interaction, *Opt. Express*, 13 (13) 5064–5073, 2005.

20. M. Pöllinger, D. O'Shea, F. Warken, and A. Rauschenbeutel, Ultrahigh-Q tunable whispering-gallery-mode microresonator, *Phys. Rev. Lett.*, 103, 053901, 2009.

21. L. P. Berenger, Perfectly matched layer for the FDTD solution of wave-structure interaction problem, *IEEE Trans. Antennas Propag.*, 44 (1) 10–118, 1996.

22. K. S. Yee, Numerical solution of initial boundary value problems involving Maxwell's equations in isotropic media, *IEEE Trans. Antennas Propagat.*, 14, 302–307, 1966.

23. M. A. Ordal, L. L. Long, R. J. Bell, S. E. Bell, R. R. Bell, R. W. Alexander Jr, and C. A. Ward, Optical properties of the metals Al, Co, Cu, Au, Fe, Pb, Ni, Pd, Pt, Ag, Ti, and W in the infrared and far infrared, *Appl. Opt.*, 22, 1099–1119, 1983.

24. W. Zhang, A. K. Azad, J. Han, J. Xu, J. Chen, and X. C. Zhang, Direct observation of a transition of a surface plasmon resonance from a photonic crystal effect, *Phys. Rev. Lett.*, 98, 183901:1–4, 2007.

25. I. Srithanachai, S. Ueamanapong, S. Niemcharoen, and P. P. Yupapin, Novel design of solar cell efficiency improvement using an embedded electron accelerator on-chip, *Optics Express*, 20 (12) 12640–12648, 2012.

4 Optical Spin

4.1 INTRODUCTION

Optical add-drop filters have been widely investigated in many areas of application, especially the use of an all-optical device which is formed with a modified add-drop optical filter known as a PANDA ring resonator [1–3]. By using an optical dark–bright soliton control arrangement within a semiconductor add-drop multiplexer, promising applications have been observed [4]. In application, one of the advantages is that the dark soliton peak signal is always at low level, which is useful for secured signal communications in the transmission link. The other applications are associated with high-optical-field configurations (such as an optical tweezer or potential well) [5]. Spin manipulation has generated tremendous research progress in many fields, such as the light-shift effect in a rubidium atom [6], magnetic clusters [7], semiconductors [8], GaAs/AlGaAs quantum well [9], carbon nanotubes [10], thin-film nanomagnets [11], magnetic tunnel junctions [12], and CdTe quantum dots [13]. All of these fields are based on pulsed magnetic resonance technique, which is well developed, but to date the single spin detection still remains a challenging task. The use of photon states for spin manipulation has become a promising technique of investigation, in which photons are the ideal candidates to transmit quantum information with little decoherency. Furthermore, spin states can be used to store and process quantum information due to their long coherence time. Therefore, the investigation of spin manipulation [14,15], spin detection, remote spin entanglement mediated by photons, and quantum-state transfer between photons and spins are of great importance. However, there is still a need to explore aspects of spin coupling and orbital motion of electrons in carbon nanotubes, magnetic manipulations, side-wall oxide effects on spin torque, magnetic field-induced reversal characteristics of thin-film nanomagnets, magnetic vortex oscillator driven by direct spin polarized current, and measurement of the spin transfer torque vector in magnetic tunnel junctions [16,17].

In this chapter, the use of a modified add-drop optical filter structure is analyzed within the PANDA ring resonator system to form the orthogonal sets of solitons, which can be decomposed into left and right circularly rotated (polarized) waves. A modified add-drop optical filter is called *PANDA* (which is a Chinese bear) by the authors. The orthogonal set of solitons can be done relative to any orthogonal set of axes, where the typical plane wave of ordinary light consists of components with all polarizations mixed together. The use of a new orthogonal set of light pulses called a *dark–bright soliton pair* [18] is proposed to form the two components of circularly rotated waves. In principle, the orthogonal solitons can be used to form the two components of polarized light, which are called *optoelectronic spin states*. Finally, the different soliton spins are described by using several pairs of orthogonal solitons, in which the conservation of their spins is maintained and realized.

4.2 THEORETICAL BACKGROUND

A dark–bright soliton conversion system using a ring resonator optical channel-dropping filter is composed of two sets of coupled waveguides, as shown in Figure 4.1. The relative phase of the two output light signals after coupling into the optical coupler is $\pi/2$. Thus, the signals coupled into the drop and through ports have acquired a phase of π with respect to the input port signal. The resonant wavelength at through port on resonance can be extinguished completely by engineering the appropriate coupling coefficients and as a result all the power would be coupled into the drop port. The control fields at the input and add ports are formed by the dark and bright optical solitons and described by Equations (4.1) and (4.2), respectively [4,5].

$$E_{in}(t) = A_0 \tanh\left[\frac{T}{T_0}\right] \exp\left[\left(\frac{z}{2L_D}\right) - i\omega_0 t\right] \tag{4.1}$$

$$E_{in}(t) = A_0 \operatorname{sech}\left[\frac{T}{T_0}\right] \exp\left[\left(\frac{z}{2L_D}\right) - i\omega_0 t\right] \tag{4.2}$$

Here A_0 and z are the optical field amplitude and propagation distance, respectively. $T = t - \beta_1 z$, where b_1 and b_2 are the coefficients of the linear and second-order terms of Taylor expansion of the propagation constant. $L_D = T_0^2/|\beta_2|$ is the dispersion length of the soliton pulse. T_0 is a soliton pulse propagation time at initial input (or soliton pulse width), where t is the soliton phase shift time, and ω_0 is the frequency shift of the soliton. The optical fields of the system can be expressed as follows

$$E_1 = -j\kappa_1 E_i + \tau_1 E_4, \tag{4.3}$$

$$E_2 = \exp(j\omega T/2)\exp(-\alpha L/4)E_1, \tag{4.4}$$

$$E_3 = \tau_2 E_2 - j\kappa_2 E_a, \tag{4.5}$$

$$E_4 = \exp(j\omega T/2)\exp(-\alpha L/4)E_3, \tag{4.6}$$

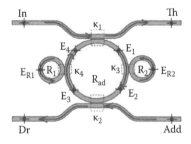

FIGURE 4.1 A schematic diagram of an InGaAsP/InP PANDA ring resonator, where R_i: ring radii, E_i: electric fields, κ_i: coupling coefficients, In: input field, Th: through port, Dr: drop port, and Add: add port.

$$E_t = \tau_1 E_i - j\kappa_1 E_4, \tag{4.7}$$

$$E_d = \tau_2 E_a - j\kappa_2 E_2, \tag{4.8}$$

Here, E_i is the input field, E_a is the add (control) field, E_t is the through field, E_d is the drop field, $E_1 \ldots E_4$ are the fields in the ring at points 1...4, κ_1 is the field coupling coefficient between the input bus and ring, κ_2 is the field coupling coefficient between the ring and output bus, L is the circumference of the ring, T is the time taken for one round trip (round-trip time), and α is the power loss in the ring per unit length. We assume that this is the lossless coupling, that is, $\tau_{1,2} = \sqrt{1 - \kappa_{1,2}^2}$. $T = Ln_{eff}/c$. The output intensities at the drop and through ports are given by

$$|E_d|^2 = \left| \frac{-\kappa_1 \kappa_2 A_{1/2} \Phi_{1/2}}{1 - \tau_1 \tau_2 A\Phi} E_i + \frac{\tau_2 - \tau_1 A\Phi}{1 - \tau_1 \tau_2 A\Phi} E_a \right|^2. \tag{4.9}$$

$$|E_t|^2 = \left| \frac{\tau_2 - \tau_1 A\Phi}{1 - \tau_1 \tau_2 A\Phi} E_i + \frac{-\kappa_1 \kappa_2 A_{1/2} \Phi_{1/2}}{1 - \tau_1 \tau_2 A\Phi} E_a \right|^2. \tag{4.10}$$

where $A_{1/2} = \exp(-\alpha L/4)$(the half-round-trip amplitude), $A = A_{1/2}^2$, $\Phi_{1/2} = \exp(j\omega T/2)$ (the half-round-trip phase contribution), and $\Phi = \Phi_{1/2}^2$.

4.3 SIMULATION RESULTS

In operation, the orthogonal soliton sets can be generated by the proposed system as shown in Figure 4.1. The optical field is fed into the PANDA system, where $R_1 = R_2 = 2.5$ μm, $R_{ad} = 30$ μm, $A_{eff} = 0.25$ μm^2, $n_{eff} = 3.14$, $n_2 = 1.3 \times 10^{-13}$ cm^2/W (for InGaAsP/InP waveguide) [19], $\alpha = 0.1$ dB/mm, gap coefficients, $\kappa_1 = \kappa_2 = 0.5$, $\kappa_3 = \kappa_4 = 0.3$, $\gamma = 0.01$, and $\lambda_0 = 1350, 1450, 1500, 1550,$ and 1600 nm. It has been observed that when the dark soliton arrays are fed into a PANDA ring, the bright soliton conversion pulses are seen at E_1 and E_4 positions (as shown in Figure 4.2a,d), and the dark soliton conversion pulses are seen at E_2 and E_3 positions (as shown in Figure 4.2b,c).

In Figure 4.2, the different wavelengths are represented by different colors, where the horizontal axis represents wavelength. The wavelength-frequency domain conversion is formed by $c = f/\lambda$ to usual time domain representation in which a bright (dark) soliton pulse repeats itself in the propagation. The dynamic soliton circulated power within a PANDA ring resonator is shown in Figures 4.3 and 4.4, with peaks (1), (2), (3), and (4) detected at positions E1, E2, E3, and E4, respectively. In the case of a large number of photons, the angular momentum of the solitons generated by a dark-soliton pump based on (port In) a PANDA ring resonator at center wavelength 1450 nm is essentially zero. This is due to the fact that half the number of photons exhibit right-hand circular motion and the other half exhibit left-hand circular motion, in which spin effect of each other due to the superposition of left and right circularly polarized solitons can be canceled, which means that the spin

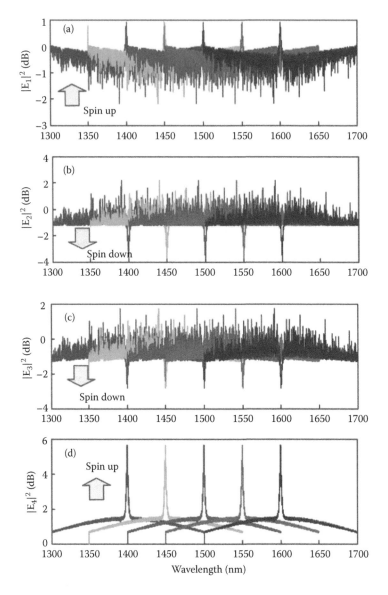

FIGURE 4.2 Many soliton spins within a PANDA ring are generated by using a dark-soliton pump input port (port *In* and where $R_1 = R_2 = 2.5\ \mu m$, $R_{ad} = 30\ \mu m$) at center wavelength 1350, 1450, 1500, 1550, and 1600 nm.

conservation is maintained. The spatial solitons are generated within a PANDA ring resonator as shown in Figure 4.4. The peak (1) illustrates a R-hand spin with power 0.2×10^{-4} dB, 0.3226 ns, peak (2) R-hand spin with power 5×10^{-3} dB, 0.5081 ns, peak (3) L-hand spin with power 8.5×10^{-3} dB, 0.7526 ns at the drop port, and peak (4) L-hand spin with power 9.5×10^{-3} dB, 0.9459 ns at the through port.

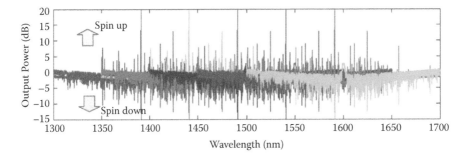

FIGURE 4.3 Many soliton spins are detected at through (spin-up) and drop (spin-down) ports of a PANDA ring resonator.

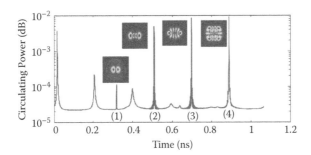

FIGURE 4.4 Many soliton spins are circulated within a PANDA ring resonator, where (1) R-hand spin detected at E_1, (2) R-hand spin detected at E_2, (3) L-hand spin detected at E_3, and (4) L-hand detected at E_4 positions, respectively.

In Figure 4.4, the spatial mode structures are shown by peaks 1, 2, 3, and 4, where they are the optical fields of the soliton pairs at the positions E1, E2, E3, and E4. The spatial mode peak 1 is the spin state of a single soliton pair after passing the first coupler, $\kappa 1$. The spatial mode peaks 2, 3, and 4 are the spin states after passing the microring, R2, second coupler, $\kappa 2$, and microring, R1, respectively.

In applications, the output signals (orthogonal solitons or photons) corresponding to the left-hand and right-hand photons can be generated and detected randomly at the same wavelength via the Th and Dr ports as shown in Figure 4.3. The angular momentum of either $+\hbar$ or $-\hbar$ is imparted to the object when a photon is absorbed by an object. Thus, the array of soliton spins can be generated and controlled by the proposed system, which is available for high-density spin states use. The spiral phase structure in a PANDA ring system causes the soliton pulse rotation for the right and left microrings faster than center ring (rad) as shown in Figures 4.5 and 4.6. In Figure 4.5, the R-hand and L-hand helical phase structures in a PANDA ring are circulated spiral rotation, in which the right and left microring signals are rotated faster than the center ring (rad) signals because the right and left microring radii (circumferences) are smaller (shorter) than the center ring.

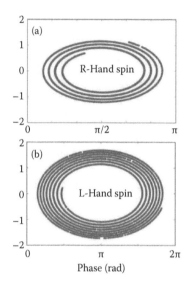

FIGURE 4.5 Helical phase structure of the orthogonal soliton pair within a PANDA ring, where (a) both within R_1 and R_2 and (b) R_{ad}.

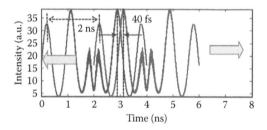

FIGURE 4.6 Dynamic soliton spins at the center wavelength, where the R-hand spin is rotated faster than the L-hand spin (i.e., 40 fs and 2 ns).

4.4 CONCLUSION

In this chapter, the new orthogonal sets are manipulated by using the orthogonal soliton pair within a PANDA ring resonator known as dark–bright soliton conversion pulses. In this manipulation, the electromagnetic radiations were generated by using the orthogonal soliton pulses. Soliton is a light pulse, which will be detected at the output ports (through or drop ports), where the energy of the soliton (light) and the photon (particle) will finally be observed by the optical detector. Hence, the spin axis of the photon is always parallel to its direction of motion. Many orthogonal sets are available, in which the spin conservation of many particles is maintained for system use on a large scale. Therefore, for future applications, the use of high-performance storage, low-power magnetic logic, quantum logic, quantum gate, nano-antenna, nanoradio, and applications of spintronic sensing can be realized based on realistic device parameters.

REFERENCES

1. T. Phatharaworamet, C. Teeka, R. Jomtarak, S. Mitatha, and P. P. Yupapin, Random binary code generation using dark-bright soliton conversion control within a PANDA ring resonator, *J. Lightw. Technol.*, 28, 2804, 2010.
2. N. Suwanpayak, M. A. Jalil, C. Teeka, J. Ali, and P. P. Yupapin, Optical vortices generated by a PANDA ring resonator for drug trapping and delivery applications, *Biomed. Opt. Express*, 2, 159, 2011.
3. P. Youplao, T. Phattaraworamet, S. Mitatha, C. Teeka, and P. P. Yupapin, Novel optical trapping tool generation and storage controlled by light, *J. Nonlin. Opt. Phys. Mater.*, 19, 371, 2010.
4. K. Sarapat, N. Sangwara, K. Srinuanjan, P. P. Yupapin, and N. Pornsuwancharoen, Novel dark-bright optical solitons conversion system and power amplification, *Opt. Eng.*, 48, 045004, 2009.
5. B. Piyatamrong, K. Kulsirirat, W. Techithdeera, S. Mitatha, and P. P. Yupapin, Dynamic potential well generation and control using double resonators incorporating in an add/drop filter, *Mod. Phys. Lett. B*, 24, 3071, 2010.
6. T. Moriyasu, D. Nomoto, Y. Koyama, Y. Fukuda, and T. Kohmoto, Spin manipulation using the light-shift effect in rubidium atoms, *Phys. Rev. Lett.*, 103, 213602, 2009.
7. W. Hübner, S. Kersten, and G. Lefkidis, Optical spin manipulation for minimal magnetic logic operations in metallic three-center magnetic clusters, *Phys. Rev. B*, 79, 184431, 2009.
8. R. C. Myers, M. H. Mikkelsen, J. M. Tang, A. C. Gossard, M. E. Flatte, and D. D. Awschalom, Zero-field optical manipulation of magnetic ions in semiconductors, *Nature Matter.*, 7, 203, 2008.
9. T. Korn, Time-resolved studies of electron and hole spin dynamics in modulation-doped GaAs/AlGaAs quantum wells, *Phys. Reports*, 494, 415, 2010.
10. F. Kuemmeth, S. Ilani, D. C. Ralph, and P. L. McEuen, Coupling of spin and orbital motion of electrons in carbon nanotubes, *Nature*, 452, 448, 2008.
11. O. Ozatay, P. G. Gowtham, K. W. Tan, J. C. Read, K. A. Mkhoyan, M. G. Thomas, G. D. Fuchs, P. M. Braganca, E. M. Ryan, K. V. Thadan, J. Silcox, D. C. Ralph, and R. A. Buhrman, Sidewall oxide effects on spin-torque- and magnetic-field-induced reversal characteristics of thin-film nanomagnets, *Nature Mater.*, 7, 567, 2008.
12. D. F. Gordon, B. Hafizi, and A. Ting, Nonlinear conversion of photon spin to photon orbital angular momentum, *Opt. Lett.*, 34, 3280, 2009.
13. M. Goryca, T. Kazimierczuk, M. Nawrocki, A. Golnik, J. A. Gaj, P. Wojnar, G. Karczewski, and P. Kossacki, Optical manipulation of a single Mn spin in a CdTe quantum dot, *Physica E*, 42, 2690, 2010.
14. R. I. Shekhter, A. M. Kadigrobov, M. Jonson, E. I. Smotrova, A. I. Nosich, and V. Korenivski, Subwavelength terahertz spin-flip laser based on a magnetic point-contact array, *Opt. Lett.*, 36, 2381, 2011.
15. V. S. Pribiag, I. N. Krivorotov, G. D. Fuchs, P. M. Braganca, O. Ozatay, J. C. Sankey, D. C. Ralph, and R. A. Buhrman, Magnetic vortex oscillator driven by d.c. spin-polarized current, *Nature Phys.*, 3, 498, 2007.
16. O. M. J. Erve, C. Awo-Affouda, A. T. Hanbicki, C. H. Li, P. E. Thompson, and B. T. Jonker, Information processing with pure spin currents in silicon: Spin injection, extraction, manipulation, and detection, *IEEE Transac. Electron Dev.*, 56, 2343, 2009.

17. E. Brasselet, Y. Izdebskaya, V. Shvedov, A. S. Desyatnikov, W. Krolikowski, and Y. S. Kivshar, Dynamics of optical spin-orbit coupling in uniaxial crystals, *Opt. Lett.*, 34, 1021, 2009.

18. Y. Zhang, Z. Wang, Z. Nie, C. Li, H. Chen, K. Lu, and M. Xiao, Four-wave mixing dipole soliton in laser-induced atomic gratings, *Phys. Rev. Lett.*, 106, 093904, 2011.

19. J. Zhu, S. K. Ozdemir, Y. F. Xiao, L. Li, L. He, D. R. Chen, L. Yang, On-chip single nanoparticle detection and sizing by mode splitting in an ultrahigh-Q microresonator, *Nat. Photon.*, 4, 46, 2010.

5 Nano-Antennas

5.1 INTRODUCTION

Nanoscale devices have played a crucial role in the nanoscopic regime, especially in information technology. One of the objectives is the search for a small device for a telephone handset, which is always the challenge. One of the key devices of the telephone handset is the antenna. Therefore, the search for a small, good performance antenna is still interesting, in which a new type of nano-antenna is recommended for modern telephone handset application. To date, the nano-antenna has been an interesting field as seen from many applications such as biology and medicine [1], monitoring and spectroscopy, medical imaging [2,3], security [3], material spectroscopy and sensing [4], and high-data-rate communications [5]. For the high frequency in THz scale, optical nano-antennas offer dominant feature capability in order to fabricate and characterize the optical antenna. Moreover, optical nano-antennas cover the potential advantages in the detection of light showing polarization, tunable, and rapid time response. Another challenge is high loss at optical frequency that becomes the limitation ability to extend current at the optical spectrum.

Photonic (optical) spin has become a very popular field nowadays; there are enormous applications for optical spin [6] such as long distance optical transport, communication network and security, quantum computers, and communication. Besides, spin mechanisms using bright and dark soliton conversion behaviors were demonstrated by Yupapin et al. [7,8], which consist of an add-drop optical filter known as a PANDA ring resonator. Soliton spin can be decomposed into left and right circularly polarized waves called a *dark–bright soliton pair*, which can be used to generate the orthogonal set of axes (transverse electric and transverse magnetic modes), in which finally, the diploe oscillation can be established. Although, there are many investigations of nano-antennas and optical spin, complete information about nano-antennas using optical spin is unavailable. Therefore, this chapter proposes a method, which reaches an accommodation between nano-antenna design and optical spin manipulation by using a PANDA ring resonator. The obtained results have shown the spin mechanism using bright and dark soliton conversion to increase the efficiency of optical nano-antenna radiation and current emission. Furthermore, the proposed system can also be used for a wide range of applications such as nanosensors, biomedicine, high-data-rate communication systems, and spectroscopy.

5.2 PHOTONIC DIPOLE

Optical dipole is formed by spin manipulation, which was explained in Glomglome et al. [9], where a ring resonator was used to control the dark–bright soliton conversion behaviors. The PANDA ring resonator was used to form the orthogonal set of a dark–bright soliton pair, which can be decomposed into right and left circularly polarized waves as shown in Figure 5.1. The relative phase of the two output light signals after coupling into the optical coupler is $\pi/2$. This means that the signals coupled into the drop port (*Dr*) and through port (*Th*) have acquired a phase of π with respect to the input port (*In*) signal. The input and control fields at the input port and control port (*Ct*) are formed by the dark and bright optical solitons as given in Glomglome et al. [9].

In operation, the orthogonal soliton sets can be generated by using the system as shown in Figure 5.2. The optical field is fed into the ring resonator system, where $R_1 = R_2 = 2.5\ \mu m$, $R_{ad} = 30\ \mu m$ by using a microring, and $R_{Th} = R_{Ct} = 20\ \mu m$. To form the initial spin states, the magnetic field is induced by an aluminum plate coupled on AlGaAs waveguides for optoelectronic spin-up and spin-down states. The coupling coefficient ratios $\kappa_1:\kappa_2$ are 50:50, 90:10, 10:90 acting in the following manner: (a) dark soliton is input into input and control ports, (b) dark and bright solitons are fed for input and control signals, (c) bright and dark solitons are used for input and control signals, and (d) bright soliton is used for input and control signals. The ring radii $R_{ad} = 30\ \mu m$, $A_{eff} = 0.25\ \mu m^2$, $n_{eff} = 3.14$ (for InGaAsP/InP) [9], $\alpha = 0.1$ dB/mm, $\gamma = 0.01$, and $\lambda_0 = 1.55\ \mu m$. Figure 5.3 shows the output intensities for spin-injected for transverse electric (TE) and transverse magnetic (TM) fields, which are generated using a PANDA ring resonator. The optoelectronic fields generated by a dark-soliton pump are based on a through port and drop port microring resonator at a center wavelength 1.45 μm. Many soliton spins are detected at through (spin up) and drop (spin down) ports of a PANDA ring resonator in Figure 5.3. The optoelectronic spin manipulation generated within a PANDA ring resonator is as shown, where the output signals are randomly obtained at the *Th* and *Dr* ports as illustrated in Figure 5.4, in which the random transverse electric (TE) and transverse magnetic (TM) fields of the solitons corresponding to the left-hand and right-hand photons can be generated and detected. The angular momentum of either $+\hbar$ or $-\hbar$ is imparted to the object when a photon is absorbed by an object, where two possible spin states known as optoelectronic spins are exhibited.

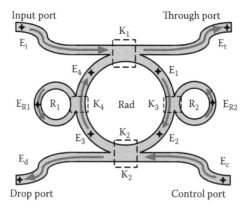

FIGURE 5.1 A schematic diagram of a PANDA ring resonator, where R_i: ring radii, E_i: electric fields, κ_i: coupling coefficients, Input field, Through port, Drop port, and Control port.

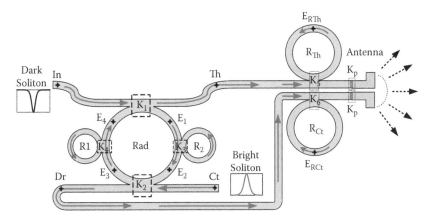

FIGURE 5.2 Schematic of an optical nano-antenna by optical spin manipulation generated within a PANDA ring resonator.

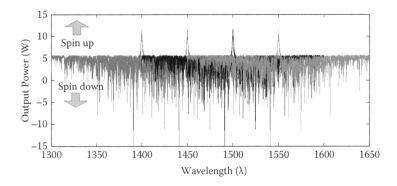

FIGURE 5.3 Many soliton spins are detected at through (spin-up) and drop (spin-down) ports. (From S. Glomglome et al., Optical Spin Generated by a Soliton Pulse in an Add-Drop Filter for Optoelectronic and Spintronic Use, *Optics & Laser Technology,* 2012, in press.)

5.3 NANO-ANTENNAS

Nano-antenna differs from radio frequency antenna in two important aspects: first, due to high losses at optical frequencies in which the assumption of perfect electrical conductor is no longer valid, and second, due to unique phenomena at nanoscale (surface plasmon polarization). The most important different response of these structures is the subwavelength field confinement. Therefore, serious efforts are being devoted to extend current understanding from radio frequency antennas to their nano-antenna counterparts. The model of an infinitesimal dipole to discuss the different characters of the fields in near-field and far-field regions is used. This model further helps to appreciate the difference between the radiated power and the power stored in the near field of an antenna. The nano-antenna system is as shown in Figure 5.5, which is used to find the optimum dipole antenna half wavelength for transmission.

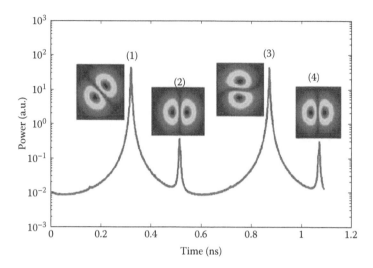

FIGURE 5.4 Photonic dipole oscillation generated within a PANDA ring resonator, where (1), (2), (3) and (4) present the different positions during oscillation within a PANDA ring. (From S. Glomglome et al., Optical Spin Generated by a Soliton Pulse in an Add-Drop Filter for Optoelectronic and Spintronic Use, *Optics & Laser Technology,* 2012, in press.)

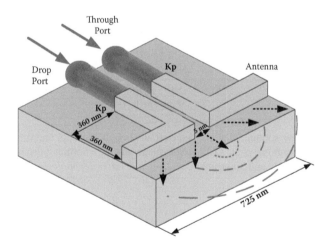

FIGURE 5.5 A nano-antenna system using optical spin.

By using the proposed system, the current density, dipole radiation, and field enhancement behavior are calculated as shown in Figure 5.6. The sinusoidal current distribution (I_S) is given by King and Harrison [10] from Equation (5.1), that I_m is the current amplitude, L the total antenna length ($\lambda/2$), a is the radius of the wire (1/500), and k is the wave vector. This first approximation is very useful to calculate the radiation patterns for antenna length L, in which the highest current density and phase are obtained.

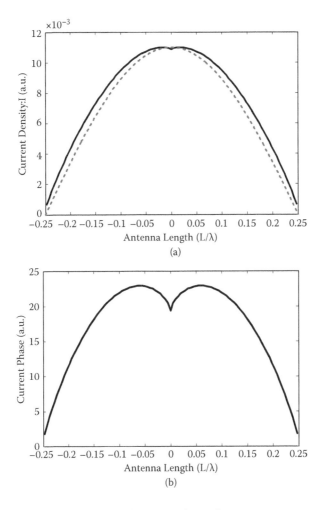

FIGURE 5.6   (a) Current density and (b) current phase of nano-antenna.

$$I_s(Z) = I_m \sin\left(k\left(\frac{1}{2}L - |Z|\right)\right) \tag{5.1}$$

To analyze the dipole antenna in more detail, a realistic current density of a dipole antenna is obtained using Pocklington's integral equation [11]. The calculation of the complex one-dimensional current density was performed by using the MATLAB functions implemented by Orfanidis [12]. The antenna input impedance is defined in Balanis [13] by the simplify impedance of the antenna for transmitted and received power at different lengths which was calculated. The input impedance for a design dipole nano-antenna with a length-to-radius ratio L/a is shown in Figure 5.7.

The optical spin performance from a PANDA ring resonator is the factor that improves the optical intensity to feed into the nano-antenna. The most

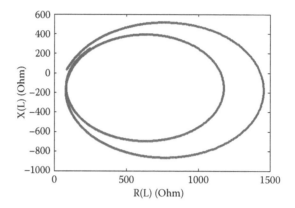

FIGURE 5.7 Impedance of nano-antenna (L/a).

popular nano-antenna material is gold, which gives a good effective resonance and dielectric constant values. The optical properties of gold follow a
Drude model, which in this work have been fitted to data in Rakic et al. [14].
The dispersion relation of the surface plasmon resonance at the interface is given
by [15]:

$$k_p = k_0 \sqrt{\frac{\varepsilon_d \varepsilon_d(\omega)}{\varepsilon_d + \varepsilon_d(\omega)}} \tag{5.2}$$

Here, k_o is the wave vector in air, ε_d is the relative permittivity of the dielectric, and $\varepsilon_m(\omega)$ is the dispersive relative permittivity of the metal. The total complex
average power delivered to the antenna is demonstrated in Equation (5.3), which
analyzed the power radiated in an infinitesimal dipole.

$$P = \frac{1}{2}|I_g|^2 Z_a = \frac{|V_g^2|}{2} \frac{R_a + X_a}{(R_a + R_g)^2 + (X_a + X_g)^2} = P_r + iP_{reac} \tag{5.3}$$

From the above equation, the imaginary part of P can be assigned to the reactive power P_{reac}, stored in the reactive near field. This time P_{reac} represents the
total reactive power. The real part of P is usually in addition to the radiate power
P_r, Z_a is the antenna impedance, I_g and V_g are the source current and voltage,
respectively.

The power radiation and field enhancement are illustrated in Figures 5.8 and
Figure 5.9, which show the power transmit from the antenna and normalized field
enhancement compared in wavelength and frequency domain. This design result
presents the optical spin method using nano-antenna that generated frequency and
radiation, which is a simple model. The optical high power source initiates the
highest surface plasmon resonance, which is vital to generate optical intensity and
polarization.

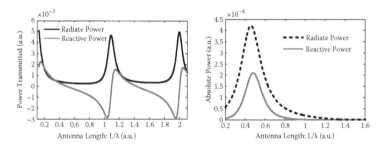

FIGURE 5.8   Power transmission of nano-antenna (L/a).

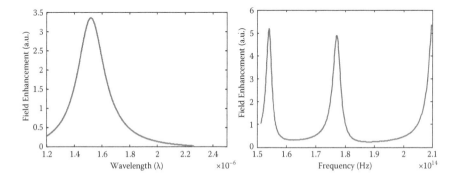

FIGURE 5.9   Nano-antenna field enhancement compared with wavelength and frequency.

5.4   CONCLUSION

This chapter proposed an interesting system and technique that can be used to form the nano-antenna. By using the photonic spin, the photonic dipole and dipole oscillation can be established, which would be available for an antenna application. Results have shown that the THz frequency generated by the two optical dipole components (TE and TM waves) of polarized light can form the orthogonal soliton pair within a PANDA ring resonator. The output signals of Th and Dr ports can form the photonic spin-up and spin-down states, which are available for wide-range nano-antenna frequency oscillation. Furthermore, our simple method can be applied for several nano-antenna designs by using an optical spin manipulation technique, which is a leading novel guideline for development in wireless nanoscale device communication. In addition, optical dipole can be used for further investigations such as dynamic dipole, dynamic torque, nanomotor, spin communication, and spin cryptography, and so forth.

REFERENCES

1. H. Altug, A. A. Yanik, R. Adato, S. Aksu, A. Artar, and M. Huang, Plasmonics for ultrasensitive biomolecular nanospectroscopy, *Optical MEMS and Nanophotonics (OPT MEMS)*, 63–64, 9–12, 2010.

2. A. Pal, A. Mehta, M. E. Marhic, K. C. Chan, and K. S. Teng, Microresonator antenna for biosensing applications, *Micro & Nano Letters, IET,* 6, 665–667, 2011.

3. C. Balocco, S. R. Kasjoo, L. Xiaofeng, Z. Linqing, Y. Alimi, S. Winnerl, B. Peng, L. Yi, L. Kin, and A. M. Song, Novel terahertz nanodevices and circuits, *10th IEEE International Conference on Solid-State and Integrated Circuit Technology (ICSICT),* Shanghai, 1176–1179, November 1–4, 2010.

4. A. Kawakami, S. Saito, and M. Hyodo, Fabrication of nano-antennas for superconducting infrared detectors, *IEEE Transactions on Applied Superconductivity,* 21, 632–635, 2011.

5. M. Bareiss, B. N. Tiwari, A. Hochmeister, G. Jegert, U. Zschieschang, H. Klauk, B. Fabel, G. Scarpa, G. Koblmuller, G. H. Bernstein, W. Porod, and P. Lugli, Nano Antenna Array for Terahertz Detection, *IEEE Transactions on Microwave Theory and Techniques,* 59, 2751–2757, 2011.

6. S. Hovel, A. Bischoff, N. C. Gerhardt, M. R. Hofmann, T. Ackemann, A. Kroner, and R. Michalzik, Optical spin manipulation of electrically pumped vertical-cavity surface-emitting lasers, *Applied Physics Letters,* 92, 041118–041118–3, 2008.

7. S. Mitatha, C. Teeka, J. Ali, and P. P. Yupapin, Soliton spin and wave-particle duality, *Optics and Photonics Letters (OPL),* 2011 (in press).

8. K. Sarapat, N. Sangwara, K. Srinuanjan, P. P. Yupapin, and N. Pornsuwancharoen, Novel dark-bright optical solitons conversion system and power amplification, *Opt. Engineering,* 48, 045004, 2009.

9. S. Glomglome, I. Srithanachai, C. Teeka, M. Mitatha, S. Niemcharoen, and P. P. Yupapin, Optical spin generated by a soliton pulse in an add-drop filter for optoelectronic and spintronic use, *Optics & Laser Technology,* 2012 (in press).

10. R. King and C. W. Jr. Harrison, The distribution of current along a symmetrical center-driven antenna, *Proceedings of the IRE,* 31, 548–567, 1943.

11. H. C. Pocklington, Electrical oscillations in wires, *Proceedings of the Cambridge Philosophical Society,* 9, 324–332, 1897.

12. S. J. Orfanidis, *Electromagnetic Waves and Antennas,* NJ: ECE Department, Rutgers University, 2004, PDF e-book, http://www.ece.rutgers.edu/~orfanidi/ewa/.

13. C. A. Balanis, Antenna theory: A review, *Proceedings of the IEEE,* 80, 7–22, 1992.

14. A. D. Rakic, A. B. Djuriic, J. M. Elazar, and M. L. Majewski, Optical properties of metallic films for vertical-cavity optoelectronic devices, *Appl. Opt.,* 37, 5271, 1998.

15. W. L. Barnes, A. Dereux, and T. W. Ebbesen, Surface plasmon subwavelength optics, *Nature,* 424, 824–830, 2003.

6 Optical Mesh Network

6.1 INTRODUCTION

Nowadays communication capacity demands in transport networks are growing rapidly. This growth is propelled mainly by an increase of multimedia traffic in the network and now further accelerated by mobile devices, smartphones, and tablets, enabling Internet access to be consumed more expediently via network connections anywhere, anytime. The advances in optical orthogonal frequency division multiplexing (OFDM) technology has enabled optical networks to compress more data through fiber optic cables by splitting carrier bandwidth frequency to multichannel carriers and optical frequency combs as compared to a regular fiber optic link. By using micro/nano devices, the design of a frequency comb generator can have new interesting aspects. However, many researchers have proposed and described the generation of optical frequency combs in various techniques [1–5] and reported the explication used for next generation communication systems in various applications such as radio-over-fiber networks with broadband wireless access [6,7], monitoring of optical intensity for in-orbit measurement response [8], stable laser development for a laser clock [9], and high speed and high spectral efficiency optical transmission for optical OFDM systems [10]. The survivability of high-capacity optical communication networks, which is important for next generation communication systems, due to interruption of service for even short duration time may be ushering in disastrous consequences, if the cost of failure is high enough. Redundancy is a common approach to improve the reliability and availability of the system. Hence, various types of redundancy mechanisms have been developed against network element failures and to increase the survivability of the network systems, for instance, study of unavailability in failure independent part protecting [11], paradigm of network segment protecting [12], graph partitioning technique for improved protection scalability in optical networks [13], and heuristic algorithms for the planning of survivable long-reach optical networks [14]. In addition to nanoscale waveguides, optical interactions and wave filtering with a nanoslit structure behavior are used [15,16], due to a very small device and guided light with very low losses, which can be useful for on-chip optical interconnects and integrated optical circuits.

The advancement of optical communication network capability and increase in capacity are needed for further improvement. In this chapter, we present a device consisting of a mesh ring resonator system that provides a fundamental three layers of high-capacity optical frequency comb for redundancy network and sensing applications. Our specific contributions in this work are:

- The design of a ring resonator mesh system can be equivalent to 6.4–8.7 GHz of frequency spacing comb lines (about 0.051–0.073 nm of free spectral range with two wavelength bands, C-band and L-band) for high-capacity optical communication and redundancy networks.

- The proposed system can be employed as a sensing system with Vernier effects for measuring the frequency shift for microscale force sensing applications.
- The proposed system can be fabricated on-chip in a few hundred micrometers with any single microwave-guide.

Section 6.2 describes the proposed system, the "Three Nodes Optical Mesh System." The simulation results of multifrequency comb generation and Vernier effects are shown in Section 6.3. The application of the proposed system, a backbone, and two redundancy networks and cross-talk effects are described in Section 6.4, and the last section contains a discussion and conclusion. However, in this article the main focus is on multifrequency comb lines generation using microrings.

6.2   THREE NODES OPTICAL MESH SYSTEM

Light from a monochromatic light source is launched into a ring resonator system with constant light field amplitude (E_0) and random phase modulation as shown in Figure 6.1, which is the combination of terms attenuation (α) and phase (ϕ_0) constants, which results in temporal coherence degradation. The time-dependent input light field (E_{in}), without pumping term, can be expressed as:

$$E_{in}(t) = E_0 \exp[-\alpha L + j\phi_0(t)] \tag{6.1}$$

where L is a propagation distance (waveguide length).

When a Gaussian pulse is input and propagated within the ring resonator system, the resonant outputs are formed. The normalized output intensity of the light fields are the ratio between the output $E_{out,n}(t)$ and input $E_{in}(t)$ fields in each round trip, which is described in the appendix of this chapter, and can be expressed as:

$$\left|\frac{E_{out1}}{E_{in}}\right|^2 = \frac{y_1\left[y_{2-5}y_{10}e^{-\frac{\alpha}{2}\frac{3L2}{2}} + 2x_{2-11} + y_{6-9}\right]\kappa3\kappa6e^{-\frac{\alpha}{2}\frac{L2}{2}}e^{-\frac{\alpha}{2}L3}}{y_6^2 y_7^2 y_{8-9} - 2x_{2-9}y_{6-7}e^{-\frac{\alpha}{2}L2}\cos(kn_2L2) + y_{2-7}e^{-\alpha L2}} \tag{6.2}$$

$$\left|\frac{E_{out2}}{E_{in}}\right|^2 = \frac{y_1 y_3 y_9 \kappa4\kappa7 e^{-\frac{\alpha}{2}\frac{L2}{2}}e^{-\frac{\alpha}{2}L4}}{y_{6-9} - 2x_{2-9}e^{-\frac{\alpha}{2}L2}\cos(kn_2L2) + y_{2-5}e^{-\alpha L2}} \tag{6.3}$$

$$\left|\frac{E_{out3}}{E_{in}}\right|^2 = \frac{y_1 y_{3-4} y_{6-9}\kappa5\kappa8 e^{-\frac{\alpha}{2}L2}e^{-\frac{\alpha}{2}L5}}{y_6^2 y_7^2 y_8^2 y_9^2 - 2x_{2-9}y_{6-9}e^{-\frac{\alpha}{2}L2}\cos(kn_2L2) + y_{2-9}e^{-\alpha L2}} \tag{6.4}$$

where $x_{a-n} = x_a x_{a+1} x_{a+2}...x_n$ and $y_{a-n} = y_a y_{a+1} y_{a+2}...y_n$, C_1 to C_{11} are constant quantities, which are expressed in Table 6.1, and the other parameters are shown in Table 6.2.

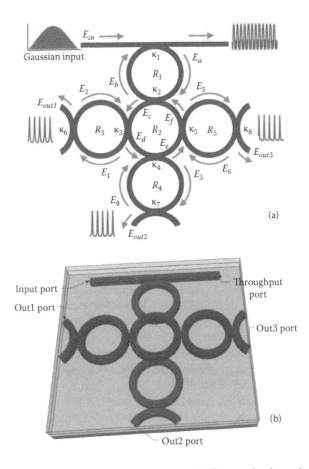

FIGURE 6.1 Optical mesh ring resonator system. (a) Proposed schematic diagram and (b) 3-D fabrication model.

TABLE 6.1
Expressed Constant Quantities: C_n

$$C_1 = \sqrt{\kappa 1}\sqrt{\kappa 2}\, e^{-\frac{\alpha}{2}\frac{L1}{2}-jkn_1\frac{L1}{2}}$$

$$C_2 = \sqrt{1-\kappa 2}-\sqrt{1-\kappa 1}\, e^{-\frac{\alpha}{2}L1-jkn_1L1}$$

$$C_3 = \sqrt{1-\kappa 3}-\sqrt{1-\kappa 6}\, e^{-\frac{\alpha}{2}L3-jkn_3L3}$$

$$C_4 = \sqrt{1-\kappa 4}-\sqrt{1-\kappa 7}\, e^{-\frac{\alpha}{2}L4-jkn_4L4}$$

$$C_5 = \sqrt{1-\kappa 5}-\sqrt{1-\kappa 8}\, e^{-\frac{\alpha}{2}L5-jkn_5L5}$$

$$C_6 = 1-\sqrt{1-\kappa 1}\sqrt{1-\kappa 2}\, e^{-\frac{\alpha}{2}L1-jkn_1L1}$$

$$C_7 = 1-\sqrt{1-\kappa 3}\sqrt{1-\kappa 6}\, e^{-\frac{\alpha}{2}L3-jkn_3L3}$$

$$C_8 = 1-\sqrt{1-\kappa 4}\sqrt{1-\kappa 7}\, e^{-\frac{\alpha}{2}L4-jkn_4L4}$$

$$C_9 = 1-\sqrt{1-\kappa 5}\sqrt{1-\kappa 8}\, e^{-\frac{\alpha}{2}L5-jkn_5L5}$$

$$C_{10} = 1-e^{-\frac{\alpha}{2}\frac{L2}{4}-jkn_2\frac{L2}{4}}$$

$$C_{11} = e^{-\frac{\alpha}{2}\frac{3L2}{4}-jkn_2\frac{3L2}{4}}$$

TABLE 6.2

Specific Variable Definition

Symbol	Quantity
κ	The coupling coefficient
$e^{\left(-\frac{\alpha}{2}L_i\right)}$	A round-trip loss coefficient
$L_i = 2\pi R_i$	A waveguide length of ring radius R_i
α	Linear absorption coefficient
$\phi_i = kn_i L_i$	The linear phase shift
$kn_i = 2\pi n_i/\lambda_i$	The propagation constant
n_i	The refractive index of waveguide length L_i
$x_n = |C_n|$	Absolute value of constant quantity C_n
$y_n = |C_n|^2$	Absolute value squared of constant quantity C_n

Equations (6.2) through (6.4) and C_1 to C_{11} indicate that the ring resonator system in this particular case is very similar to a Fabry–Perot cavity, which has an input and output mirror with a field reflectivity, (1–κ), and a fully reflecting mirror. κ is the coupling coefficient. The input optical field, E_{in}, as shown in Equation (6.1), that is, a Gaussian pulse, is input into the ring resonator system. By using the appropriate parameters, the output 1, output 2, and output 3 signals are obtained by using Equations (6.2), (6.3), and (6.4), where E_{out1}, E_{out2}, and E_{out3} represent the optical fields of the output 1, output 2, and output 3 ports, respectively. The transmitted output can be controlled by choosing the suitable coupling ratio and the radius of the ring resonators. The waveguide (ring resonator system) loss is $\alpha = 0.5$ dBmm^{-1} and the fractional coupler intensity loss is $\gamma = 0.1$.

From Figure 6.1, in principle, light pulse is sliced to the discrete signal and propagated within the first R_1 ring, and spread to other rings, R_2 to R_5, with the direction as shown in Figure 6.1a. Finally, the required signals can be obtained via output 1, output 2, and output 3 ports of the ring resonator system. In operation, an optical field as Gaussian pulse from a laser source at the specified frequency ranges (wavelength) is input into the system.

6.3 MULTIFREQUENCY COMB GENERATION

6.3.1 FREQUENCY COMBS FOR NETWORK REDUNDANCY

From Figure 6.2, the Gaussian pulse with specified frequency ranges from 192100 to 195200 GHz (1560.61 to 1535.82 nm, C-band), about 25 nm bandwidth, peak power at 2 W, is input into the mesh ring resonator system as shown in Figure 6.2a. The large bandwidth signals can be seen at the throughput port, whereas the output 1 port, output 2 port, and output 3 port of the system are shown in Figure 6.2b,c,d,e and the enlarged frequency axes of output 1, output 2, and output 3 in Figure 6.2g,h,j,k, respectively. The ring parameters, ring radii $R_1 = 420$ μm, $R_2 = 420$ μm, $R_3 = 420$ μm,

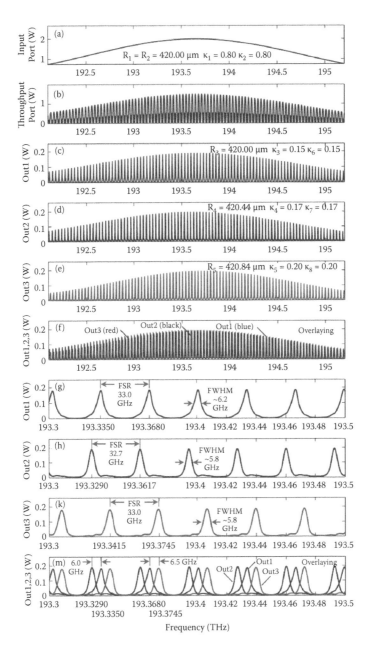

FIGURE 6.2 Simulation results for Gaussian input at frequency ranges from 192100 to 195200 GHz, C-band (1560.61 to 1535.82 nm).

$R_4 = 400.44 \, \mu m$, and $R_5 = 400.84 \, \mu m$ are used. In order to associate the system with the practical device [17], the selected parameters of the system are fixed to $n = 3.47$ (*Si-Crystalline silicon*), $\alpha = 0.5 \, dBmm^{-1}$, $\gamma = 0.1$. In this investigation, the coupling coefficient, κ of the ring resonator, are ranged from 0.15 to 0.80. The three output signals (from output 1, output 2, and output 3 ports) are shown in Figure 6.2f,m, where it is found that the system can be equivalent to 6.0–6.5 GHz frequency spacing of frequency comb lines (about 0.048–0.052 nm), $90 \times 3 = 270$ channels, which is 25 nm of input signal bandwidth. The maximum output power obtained is 0.18 W, in which the Vernier effects can be neglected.

From Figure 6.3, the Gaussian pulse with specified frequency ranges from 187500 to 190600 GHz (1598.89 to 1572.89 nm, *L*-band), peak power at 2 W is input into the mesh ring resonator system as shown in Figure 6.3a. The bandwidth signals can be seen at the output 1 port, output 2 port, and output 3 port of the mesh ring system as shown in Figure 6.3b,c,d, respectively. The ring parameters, ring radii $R_1 = 420 \, \mu m$, $R_2 = 420 \, \mu m$, $R_3 = 420 \, \mu m$, $R_4 = 400.20 \, \mu m$, and $R_5 = 400.60 \, \mu m$ are used and the selected parameters of the system are fixed as $n = 3.47$, $\alpha = 0.5 \, dBmm^{-1}$, and $\gamma = 0.1$.

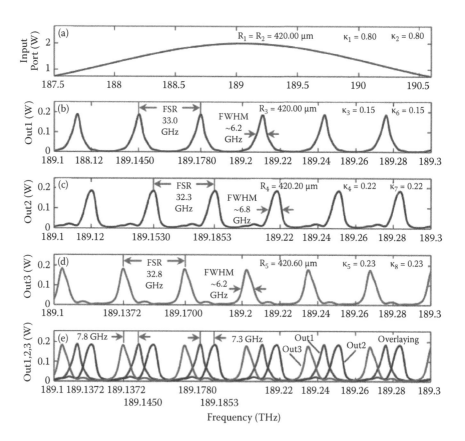

FIGURE 6.3 Simulation results for Gaussian input at frequency ranges from 187500 to 190600 GHz, *L*-band (1598.89 to 1572.89 nm).

The coupling coefficient, κ of the ring resonator, are ranged from 0.15 to 0.80. The three output signals (from output 1, output 2, and output 3 ports) are as shown in Figure 6.3e, where the system can be equivalent to 7.3–7.8 GHz frequency spacing of frequency comb lines (about 0.061–0.065 nm of FSR), $90 \times 3 = 270$ channels, with merely 25 nm of input signal bandwidth. The maximum output power obtained is 0.20 W.

6.3.2 MULTISENSING WITH VERNIER EFFECTS

An optical resonant structure such as a microring resonator is commonly used to generate the optical Vernier effects, which can be used for high-resolution distance measurement. In this work, we propose a mesh ring device to measure a force in microscale by measuring the spacing shift among the sensors and reference frequencies. The schematic diagram of a microforce sensing system using a mesh ring resonator is shown in Figure 6.4, with the same parameters as mentioned above, except the ring radii $R_1 = 500$ µm; $R_2 = 700$ µm; the reference ring $R_3 = 600$ µm; the first sensing ring $R_4 = 200$ µm, and the second sensing ring $R_5 = 400$ µm. In operation, the radius of both sensing rings is changed by the shifted signals in the interferometer ring (R_1 and R_2), where the change in optical path length is related to the change of external parameters, which can be measured. By using the proposed system, the phase shift resolution within the nanoscale region can be obtained. For instance, the exert force can be performed as a diffused force within thin film, which is coated on the sensing units, in which the optical path length is also changed, where finally the sensing and reference signals are observed and compared. The concept of stress and strain on the sensing device by the elastic modulus of the materials, which is caused by the difference in the peak spectrum of signals, is described by (6.5).

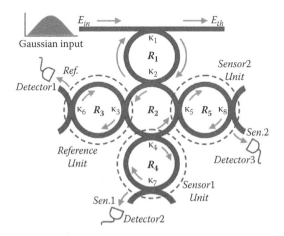

FIGURE 6.4 Schematic diagram of microforce sensing system using the mesh ring resonator.

$$Y_0 = \frac{Stress}{Strain} = \left(\frac{F}{A} \times \frac{L}{\Delta L} \right) \tag{6.5}$$

The relationship between the force and the change in sensing device length is described by

$$F = \left(\frac{Y_0 A_0}{L_0} \right) \times \Delta L \tag{6.6}$$

where F is the exerted force, Y_0 is the Young modulus, A_0 is the initial cross-sectional area of the sensing waveguide, L_0 is the initial waveguide length, and ΔL is the change of waveguide length, and is well described by the published work by Sirawattananon et al. [18].

6.4   FREQUENCY ROUTER AND CROSS-TALK EFFECTS

Generally, a frequency router (identical to wavelength router) is made up of add/drop filters; its performance depends on add/drop filters. (See Figures 6.5, 6.6, and 6.7.) To date, the nano waveguides are gaining prominence. Filters offer good

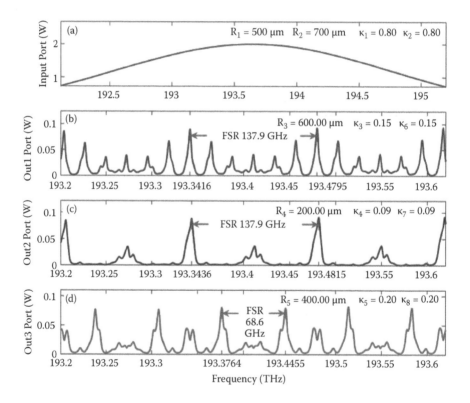

FIGURE 6.5   Simulation results of Vernier effects for Gaussian input at frequency ranges from 192100 to 195200 GHz (1560.61 to 1535.82 nm).

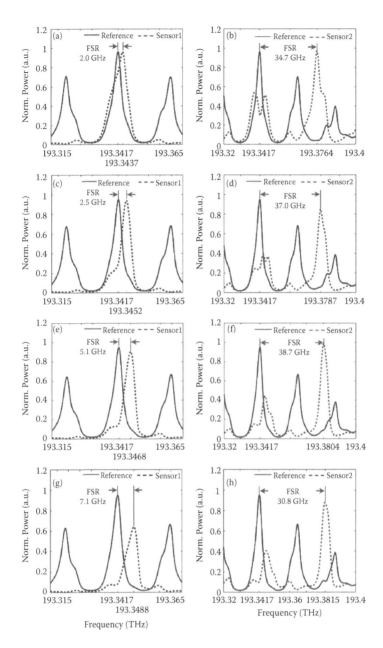

FIGURE 6.6 Spectrums of sensor units are compared with reference units, (a) and (b); before exert force response, (c)–(d), (e)–(f), and (g)–(h) are the response of exert force testing with 100, 200, and 300 μN, respectively.

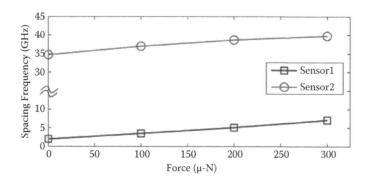

FIGURE 6.7 Linearity relation of the force and spacing shift of reference and sensor frequency.

stability and isolation between channels at moderate cost. For one thing, the add/drop filter capability will affect the size of the network. The maximum nodes of a network depend on the amount of total channels of add/drop filter. The popular DWDM product has 40 channels [17,19] and can easily build a feasible quantum key distribution network [20]. Consider the case where the multifrequency channel communication is not as popular as the Internet; this net size may be enough for a big city and of sufficient availability for future users. Second, the insertion loss will reduce the efficiency of multichannel communication distance. Since optical signals will pass through the add/drop filter, when they pass the router, the insertion loss of popular product is 5 dB. Along with the development of wavelength division multiplex (WDM) and OFDM technology, the insertion loss will be less than 1 dB in the future. The third problem is cross-talk. For a network, cross-talk will bring bit errors, so it must be reduced to as low as possible. Consider the situation that the probability of input optical signals is not the same. A terrible situation is that input optical signals which produce efficient signals pass through the waveguide that has X dB insertion loss before passing through the frequency router but those optical signals which produce cross-talk do not have insertion loss. Where $FC_j(f_i)$ is cross-talk of optical signals with frequency f_i which exports from port j, the cross-talk versus efficient signals [20] will become:

$$\text{Cross-talk versus efficient signals} = 10^{\frac{X+2FC_j(f_i)}{10}} \tag{6.7}$$

If there are many inputs that produce cross-talk, the ratio must be:

$$\text{Cross-talk versus efficient signals} = \sum_{j=1}^{N-1} 10^{\frac{X+2FC_j(f_i)}{10}} \tag{6.8}$$

Where $j \neq i$, here N is the amount of channels. In this work, $N = 90$, $X < 10$, $FC_j(f_i) < -25$ dB (where $j = I \pm 1$), $FC_j(f_i) < -30$ dB (where $j \neq I \pm 1$). The ratio is less than 0.107%.

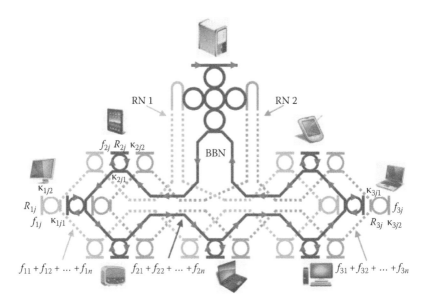

$f_{11}+f_{12}+ \cdots +f_{1n}$ $f_{21}+f_{22}+ \cdots +f_{2n}$ $f_{31}+f_{32}+ \cdots +f_{3n}$

FIGURE 6.8 System of frequency router with redundancy networks, available for backbone network instead (failure or maintenance), where BBN: backbone network, RN: redundancy network, R_{xj}: ring radii, f_{xj}: output frequency, and κ_{xji} are coupling coefficients.

So the errors caused by cross-talk are less than other effects and can be ignored (Figure 6.8). Along with the development of WDM technology, cross-talk will be smaller and the performance of the frequency router will improve. In this chapter, we provide a new scheme for a three-layer multifrequency comb network. The kernel part of this network is a frequency router, which is made up of a mesh ring resonator system. Based on the present OFDM and optical frequency comb technology, we can easily build a feasible optical frequency network with two redundancy networks on a 25 nm bandwidth.

6.5   DISCUSSION AND CONCLUSION

In this investigation, the mesh ring resonator system parameters, such as ring radius and coupling coefficients, are studied by simulation. In practice, maybe the actual fabricated device has unavoidable errors from the fabrication process and this can affect the designing response. However, the resonant frequency response of a microring resonator can be tuned in several ways. The most straightforward approach is to change the effective index of the ring, such as the thermo-optic effect; by applying heat to the ring, the refractive index of the material changes [21], and the electro-optic effect causes the change in refractive index [22]. The carrier injection: optical pumping creates free carriers (single-photon or two-photon absorption), which change the loss parameter and the refractive index of the material [23,24]. Therefore, the fabrication work needs to be considered with additional waveguide tuning technique in order to function properly.

In this chapter, we have described the new mesh ring resonator system of multi-frequency comb generation which can be equivalent to 6.4–8.7 GHz of frequency spacing comb lines (about 0.051–0.073 nm of free spectral range) with two wavelength bands, C-band and L-band, 90 × 3 = 270 channels, for high capacity used in optical networking technologies and application as an on-microchip device for redundancy optical network systems. The simulation results indicate that the proposed system can increase the network's capability and reliability, and minimize unavailability for the next generation optical communication networks.

Furthermore, the proposed mesh ring system can be employed as microscale sensing applications with multisensor units based on Vernier effects, such as microforce sensors with a few hundred μ-N, by measuring the spacing of frequency shift. The applied force can be performed as a distributed force within thin-film material, which is coated on the sensing unit. The calibration is allowed by varying frequency response of sensing and reference signals, which exist within the system and can be calibrated by comparing the measurement of sensing and reference signals without an additional optical unit. However, in this research work, the main emphasis is to generate and simulate the multifrequency comb bands.

APPENDIX

From Figure 6.1a, the relationship of the electric fields E_1–E_c, E_2–E_c, E_3–E_d, E_4–E_d, E_5–E_e, and E_6–E_e can be expressed as:

$$E_1 = \frac{E_c j\sqrt{\kappa 3}\, e^{-\frac{\alpha}{2}\frac{L2}{4}-jkn_2\frac{L2}{4}}}{1-\sqrt{1-\kappa 3}\sqrt{1-\kappa 6}\, e^{-\frac{\alpha}{2}L3-jkn_3 L3}} \tag{6.9}$$

$$E_2 = \frac{E_c j\sqrt{\kappa 3}\, e^{-\frac{\alpha}{2}\frac{L2}{4}-jkn_2\frac{L2}{4}}\sqrt{1-\kappa 6}\, e^{-\frac{\alpha}{2}\frac{L3}{2}-jkn_3\frac{L3}{2}}}{1-\sqrt{1-\kappa 3}\sqrt{1-\kappa 6}\, e^{-\frac{\alpha}{2}L3-jkn_3 L3}} \tag{6.10}$$

$$E_3 = \frac{E_d j\sqrt{\kappa 4}\, e^{-\frac{\alpha}{2}\frac{L2}{4}-jkn_2\frac{L2}{4}}}{1-\sqrt{1-\kappa 7}\sqrt{1-\kappa 4}\, e^{-\frac{\alpha}{2}L4-jkn_4 L4}} \tag{6.11}$$

$$E_4 = \frac{E_d j\sqrt{\kappa 4}\, e^{-\frac{\alpha}{2}\frac{L2}{4}-jkn_2\frac{L2}{4}}\sqrt{1-\kappa 7}\, e^{-\frac{\alpha}{2}\frac{L4}{2}-jkn_4\frac{L4}{2}}}{1-\sqrt{1-\kappa 7}\sqrt{1-\kappa 4}\, e^{-\frac{\alpha}{2}L4-jkn_4 L4}} \tag{6.12}$$

$$E_5 = \frac{E_e j\sqrt{\kappa 5}\, e^{-\frac{\alpha}{2}\frac{L2}{4}-jkn_2\frac{L2}{4}}}{1-\sqrt{1-\kappa 8}\sqrt{1-\kappa 5}\, e^{-\frac{\alpha}{2}L5-jkn_5 L5}} \tag{6.13}$$

$$E_6 = \frac{E_e j\sqrt{\kappa 5}\, e^{\frac{\alpha}{2}\frac{L2}{4}-jkn2\frac{L2}{4}}\sqrt{1-\kappa 8}\, e^{\frac{\alpha}{2}\frac{L5}{2}-jkn5\frac{L5}{2}}}{1-\sqrt{1-\kappa 8}\sqrt{1-\kappa 5}\, e^{-\frac{\alpha}{2}L5-jkn5 L5}}$$ (6.14)

The quantities C_n are defined in Table 6.1; the electric fields E_a–E_f can be expressed as:

$$E_a = \frac{E_f j\sqrt{\kappa 2}\, e^{\frac{\alpha}{2}\frac{L2}{4}-jkn2\frac{L2}{4}}\sqrt{1-\kappa 1}\, e^{\frac{\alpha}{2}\frac{L1}{2}-jkn1\frac{L1}{2}}+E_{in} j\sqrt{\kappa 1}}{1-\sqrt{1-\kappa 1}\sqrt{1-\kappa 2}\, e^{-\frac{\alpha}{2}L1-jkn1 L1}}$$ (6.15)

$$E_b = \frac{E_f j\sqrt{\kappa 2}\, e^{\frac{\alpha}{2}\frac{L2}{4}-jkn2\frac{L2}{4}}+E_{in} j\sqrt{\kappa 1}\sqrt{1-\kappa 2}\, e^{\frac{\alpha}{2}\frac{L1}{2}-jkn1\frac{L1}{2}}}{1-\sqrt{1-\kappa 1}\sqrt{1-\kappa 2}\, e^{-\frac{\alpha}{2}L1-jkn1 L1}}$$ (6.16)

$$E_c = \frac{E_f C_2\, e^{-\frac{\alpha}{2}\frac{L2}{4}-jkn2\frac{L2}{4}}-E_{in}C_1}{C_6}$$ (6.17)

$$E_d = \frac{\left(E_f C_2 e^{\frac{\alpha}{2}\frac{L2}{4}-jkn2\frac{L2}{2}}-E_{in}C_1\right)C_3}{C_6 C_7}$$ (6.18)

$$E_e = \frac{\left(E_f C_2 e^{\frac{\alpha}{2}\frac{L2}{2}-jkn2\frac{L2}{2}}-E_{in}C_1\right)C_3 C_4 e^{\frac{\alpha}{2}\frac{L2}{4}-jkn2\frac{L2}{4}}}{C_6 C_7 C_8}$$ (6.19)

$$E_f = \frac{E_{in}C_1 C_3 C_4 C_5\, e^{-\frac{\alpha}{2}\frac{L2}{2}-jkn2\frac{L2}{2}}}{C_2 C_3 C_4 C_5\, e^{-\frac{\alpha}{2}L2-jkn2 L2}-C_6 C_7 C_8 C_9}$$ (6.20)

where the electric fields at out1, out2, and out3 port are given by:

$$E_{out1} = E_1 j\sqrt{\kappa 6}\, e^{-\frac{\alpha}{2}\frac{L3}{2}-jkn3\frac{L3}{2}}$$ (6.21)

$$E_{out2} = E_3 j\sqrt{\kappa 7}\, e^{-\frac{\alpha}{2}\frac{L4}{2}-jkn4\frac{L4}{2}}$$ (6.22)

$$E_{out3} = E_5 j\sqrt{\kappa 8}\, e^{-\frac{\alpha}{2}\frac{L5}{2}-jkn5\frac{L5}{2}}$$ (6.23)

thus, we can express the electric fields E_{out1}, E_{out2}, and E_{out3} as:

$$E_{out1} = \frac{E_{in}C_1(C_2C_3C_4C_5C_{10}C_{11}+C_6C_7C_8C_9)\sqrt{\kappa3}\sqrt{\kappa6}\,e^{-\frac{\alpha}{2}\frac{L2}{4}-jkn_2\frac{L2}{4}}\,e^{-\frac{\alpha}{2}\frac{L3}{2}-jkn_3\frac{L3}{2}}}{C_6^2C_7^2C_8C_9 - C_2C_3C_4C_5C_6C_7\,e^{-\frac{\alpha}{2}L2-jkn_2L2}}$$

$$E_{out2} = \frac{E_{in}C_1C_3C_9\sqrt{\kappa4}\sqrt{\kappa7}\,e^{-\frac{\alpha}{2}\frac{L2}{4}-jkn_2\frac{L2}{4}}\,e^{-\frac{\alpha}{2}\frac{L4}{2}-jkn_4\frac{L4}{2}}}{C_6C_7C_8C_9 - C_2C_3C_4C_5\,e^{-\frac{\alpha}{2}L2-jkn_2L2}}$$

$$E_{out3} = \frac{E_{in}C_1C_3C_4C_6C_7C_8C_9\sqrt{\kappa5}\sqrt{\kappa8}\,e^{-\frac{\alpha}{2}\frac{L2}{2}-jkn_2\frac{L2}{2}}\,e^{-\frac{\alpha}{2}\frac{L5}{2}-jkn_5\frac{L5}{2}}}{C_6^2C_7^2C_8^2C_9^2 - C_2C_3C_4C_5C_6C_7C_8C_9\,e^{-\frac{\alpha}{2}L2-jkn_2L2}}$$

Finally, output intensity of the light fields are the ratio between the output and input fields in each round trip which can be expressed as (6.2) through (6.4).

REFERENCES

1. C. Villa, M. Hayduk, and E. Donkor, Terahertz optical frequency comb generation by spectral filtering of broadband spontaneous amplified emissions from a semiconductor optical amplifier, *Journal of Lightwave Technology*, 27 (23) 5437–5441, 2009.
2. N. Dupuis, C. R. Doerr, L. Zhang, L. Chen, N. J. Sauer, P. Dong, L. L. Buhl, and D. Ahn, InP-Based comb generator for optical OFDM, *Journal of Lightwave Technology*, 30 (4) 466–472, 2012.
3. Y. Dou, H. Zhang, and M. Yao, Generation of flat optical-frequency comb using cascaded intensity and phase modulators, *IEEE Photonics Technology Letters*, 24 (9) 727–729, 2012.
4. E. Sooudi, S. Sygletos, A. D. Ellis, G. Huyet, J. G. McInerney, F. Lelarge, K. Merghem, R. Rosales, A. Martinez, A. D. Ramdane, and S. P. Hegarty, Optical frequency comb generation using dual-mode injection-locking of quantum-dash mode-locked lasers: Properties and applications, *IEEE Journal of Quantum Electronics*, 48 (10) 1327–1337, 2012.
5. P. J. Delfyett, I. Ozdur, N. Hoghooghi, M. Akbulut, J. D. Rodriguez, and S. Bhooplapur, Advanced ultrafast technologies based on optical frequency combs, *IEEE Journal of Selected Topics in Quantum Electronics*, 18 (1) 258–274, 2012.
6. A. Chowdhury, H. C. Chien, Y. T. Hsueh, and G. K. Chang, Advanced system technologies and field demonstration for in-building optical-wireless network with integrated broadband services, *Journal of Lightwave Technology*, 27 (12) 1920–1927, 2009.
7. X. Xie, Y. Dai, Y. Ji, K. Xu, Y. Li, J. Wu, and J. Lin, Broadband photonic radio-frequency channelization based on a 39-GHz optical frequency comb, *IEEE Photonics Technology Letters*, 24 (8) 661–663, 2012.
8. S. Mizobuchi, K. Kikuchi, S. Ochiai, T. Nishibori, T. Sano, K. Tamaki, and H. Ozeki, In-orbit measurement of the AOS (acousto-optical spectrometer) response using frequency comb signals, *IEEE Journal of Selected Topics in Applied Earth Observations and Remote Sensing*, 5 (3) 977–983, 2012.

9. K. Hosaka, H. Inaba, Y. Nakajima, M. Yasuda, T. Kohno, A. Onae, and F. L. Hong, Evaluation of the clock laser for an Yb lattice clock using an optical fiber comb, *IEEE Transactions on Ultrasonics, Ferroelectrics, and Frequency Control*, 57 (3) 606–612, 2010.

10. X. Yi, N. K. Fontaine, R. P. Scott, and S. J. Ben Yoo, Tb/s coherent optical OFDM systems enabled by optical frequency combs, *Journal of Lightwave Technology*, 28 (14) 2054–2061, 2010.

11. A. Ranjbar and C. Assi, Availability-aware design in mesh networks with failure-independent path-protecting p-cycles, *IEEE Transactions on Reliability*, 58 (2) 348–363, 2009.

12. D. P. Onguetou and W. D. Grover, A two-hop segment protecting paradigm that unifies node and span failure recovery under p-cycles, *IEEE Communications Letters*, 14 (11) 1080–1082, 2010.

13. H. Drid, B. Cousin, M. Molnar, and N. Ghani, Graph partitioning for survivability in multi-domain optical networks, *IEEE Communications Letters*, 14 (10) 978–980, 2010.

14. B. Kantarci and H. T. Mouftah, Availability and cost-constrained long-reach passive optical network planning, *IEEE Transactions on Reliability*, 61 (1) 113–124, 2012.

15. J. Wuenschell and H. K. Kim, Excitation and propagation of surface plasmons in a metallic nanoslit structure, *IEEE Transactions on Nanotechnology*, 7 (2) 229–236, 2008.

16. B. Guo, G. Song, and L. Chen, Resonant enhanced wave filter and waveguide via surface plasmons, *IEEE Transactions on Nanotechnology*, 8 (2) 408–411, 2009.

17. H. Takara, T. Ohara, T. Yamamoto, H. Masuda, M. Abe, H. Takahashi, and T. Morioka, Field demonstration of over 1000-channel DWDM transmission with supercontinuum multi-carrier source, *Electronics Letters*, 41 (5) 270–271, 2002.

18. C. Sirawattananon, M. Bahadoran, J. Ali, S. Mitatha, and P. P. Yupapin, Analytical vernier effects of a PANDA ring resonator for microforce sensing application, *IEEE Transactions on Nanotechnology*, 11 (4) 707–712, 2012.

19. K. Sarapat, J. Ali, and P.P. Yupapin, A novel storage and tunable light source generated by a soliton pulse in a micro ring resonator system for super dense wavelength division multiplexing use, *Microwave and Optical Technology Letters*, 51 (12) 2948–2952, 2009.

20. P. Youplao, P. Pongwongtragull, S. Mitatha, and P. P. Yupapin, Crosstalk effects of quantum key distribution via a quantum router, *Microwave and Optical Technology Letters*, 53 (5) 1094–1099, 2011.

21. M. B. J. Diemeer, Polymeric thermo-optic space switches for optical communications, *Optical Materials*, 9, 192–200, 1998.

22. I. L. Gheorma and R. M. Osgood, Fundamental limitations of optical resonator based high-speed EO modulators, *IEEE Photonics Technology Letters*, 14 (6) 795–797, 2002.

23. T. A. Ibrahim, W. Cao, Y. Kim, J. Li, J. Goldhar, P. T. Ho, and C. H. Lee, Alloptical switching in a laterally coupled microring resonator by carrier injection, *IEEE Photonics Technology Letters*, 15 (1) 36–38, 2003.

24. V. R. Almeida, C. A. Barrios, R. R. Panepucci, and M. Lipson, All-optical control of light on a silicon chip, *Nature*, 431, 1081–1084, 2004.

7 Micro-Optical Gyroscope

7.1 INTRODUCTION

Micro-optical gyroscope (MOG) has advantages due to integrated optoelectronic technology, where the ring resonator can be fabricated by optical integrated technique. The MOG is a novel type of angle-velocity sensor, which is small, highly precise, and low cost, and so forth [1,2]. Recently, the three-axis gyroscope has been proposed for measurement of three-axis angular velocity [3,4] based on the inertial navigation systems for various applications. The compensation of gyro error has improved the inertial navigation system (INS), which is an important development in calibration technology [5,6]. The general network architecture, such as star, ring, and bus topology can access from a central control (CC) to subscriber. The ring resonator has been used for various applications [7–10], such as dark–bright soliton conversion, random binary code, tweezers generation, and nanosensing applications. In this chapter, a single ring resonator with two lateral nonlinear ring resonators, which is called a *modified add-drop filter* (a PANDA ring resonator) is used for the fine adjustment of the micro-optical gyroscope. A new system of MOG network using a micro-embedded system consisting of a PANDA ring resonator and CC is proposed. The star topology is an attractive network [11,12]. Modified star-ring architecture with a self-healing function is used to increase the reliability of the sensing network, where each node can perform three-dimensional measurement of rotation motion. The self-calibration among the four-point probe can also be realized using measurement data, that is, time delay of the four probe signals and each of the side rings. Furthermore, by using the central control, the adjustment and control can be performed for the real-time measurement data in any situation. In application, this technique can be useful for distributed sensors, in which the self-calibration is available, where the fine adjustment of the rotational motion is required.

7.2 PRINCIPLE

The angle misalignments between the gyro input axis and the initial navigation system's body axis cause the measurement errors. The measuring coordinate system of the optical gyroscope unit can be expressed by $ox^g\, y^g\, z^g$ and body axes in the coordinate system can be expressed by $ox^b\, y^b\, z^b$. In the ideal case, the gyroscope output (N_g) has no scale factor, bias, and installation errors. The angular velocity input from the gyroscope output is given as

$$\omega_{ib}^b = K \cdot N_g \tag{7.1}$$

The actual angular velocity with errors is given as

$$\omega_{ib}^b = b_g + C_g^b K_g N_g + \delta\varepsilon^b \tag{7.2}$$

The gyroscope output error is expressed by

$$\varepsilon^b = b_g + (S_g + \Delta C_g^b)\omega_{ib}^b + \delta\varepsilon^b \tag{7.3}$$

Here, b_g is gyro bias error, S_g is gyro scale factor error, ΔC_g^b is gyro installation error, and $\delta\varepsilon^b$ is the gyro random drift.

Equation (7.3) can be transformed to be a matrix as

$$
\begin{bmatrix} \varepsilon_x^b \\ \varepsilon_y^b \\ \varepsilon_z^b \end{bmatrix} = \begin{bmatrix} b_{gx} \\ b_{gy} \\ b_{gz} \end{bmatrix} + \left(\begin{bmatrix} S_{gx} & 0 & 0 \\ 0 & S_{gy} & 0 \\ 0 & 0 & S_{gz} \end{bmatrix} + \begin{bmatrix} 0 & E_{xz} & -E_{xy} \\ -E_{yz} & 0 & E_{yx} \\ E_{zy} & -E_{zx} & 0 \end{bmatrix} \right)
$$

$$
\times \begin{bmatrix} \omega_x^b \\ \omega_y^b \\ \omega_z^b \end{bmatrix} + \begin{bmatrix} \delta\varepsilon_x^b \\ \delta\varepsilon_y^b \\ \delta\varepsilon_z^b \end{bmatrix} \tag{7.4}
$$

where $\varepsilon_x^b, \varepsilon_y^b$, and ε_z^b are defined in Equation (7.5) as

$$\varepsilon_x^b = b_{gx} + S_{gx}\omega_x^b + E_{xz}\omega_y^b - E_{xy}\omega_z^b + \delta\varepsilon_x^b$$

$$\varepsilon_y^b = b_{gy} + S_{gy}\omega_y^b - E_{yz}\omega_x^b + E_{yx}\omega_z^b + \delta\varepsilon_y^b \tag{7.5}$$

$$\varepsilon_z^b = b_{gz} + S_{gz}\omega_z^b + E_{zy}\omega_x^b - E_{zx}\omega_y^b + \delta\varepsilon_z^b.$$

Figure 7.1 shows a three-axis gyroscope structure using a PANDA ring resonator, which consists of three microring resonators; first and second ring radii with $R_l = R_r = 1.550\ \mu m$ are set as the reference and sensing units, respectively. The third ring $R_c = 3.10\ \mu m$ is formed to be an interferometer, which is located at the center of the system, and the coupler coefficients of left (R_l), right (R_r), and center rings (R_c) are $\kappa_l = 0.97$, $\kappa_r = 0.97$, and $\kappa_1 = \kappa_2 = 0.99$, respectively. The variation in rotating angular velocity and horizontal velocity with difference in phase causes the time difference (ΔT) due to the phase shift (delay time) after rotation, which can be observed and measured.

In operation, the two identical monochromatic pulses with optical field of (E_{in}) are introduced into the system at the input port 1 and input port 2, respectively, which is given by

$$E_{in}(t) = E_0 \exp[-\alpha L + j\varphi_0(t)] \tag{7.6}$$

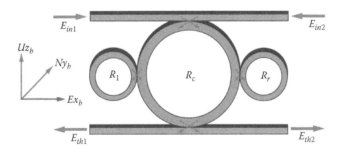

FIGURE 7.1 A schematic of a four-point probe gyroscope system using a PANDA ring resonator.

Here, $L = 2\pi R$ is a propagation distance (waveguide length), α is an attenuation, ϕ_0 are the phase constants, and L is an optical path length.

The normalized output of the light field is the ratio between the output and input fields $[E_{out}(t)$ and $E_{in}(t)]$ in each round trip given by

$$\left|\frac{E_{out}(t)}{E_{in}(t)}\right|^2 = (1-\gamma)\left[1-\frac{(1-(1-\gamma)x^2)\kappa}{\left(1-x\sqrt{1-\gamma}\sqrt{1-\kappa}\right)^2+4x\sqrt{1-\gamma}\sqrt{1-\kappa}\sin^2\left(\frac{\varphi}{2}\right)}\right] \quad (7.7)$$

When the system rotates with the angular velocity Ω_p, the phase shift (delay time) at output 1 and output 2 can be measured, which refers to the Sagnac effects [13,14] and equation of the three-axis gyroscope.

The normalized intensity of the output light field, which is the ratio between the absolute square of the throughput port and input port fields, are described by the following equations:

$$\left|\frac{E_{th1}}{E_{in1}}\right|^2 = \left|\beta_r\beta_1\beta_2 Z_c^2\right|^2 \quad (7.8)$$

$$\left|\frac{E_{th2}}{E_{in2}}\right|^2 = \left|\beta_l\beta_1\beta_2 Z_c^2\right|^2 \quad (7.9)$$

Here, β_1, β_2, β_l, and β_r are complex coefficients [15], where the authors have analyzed and obtained the precise signal at the phase difference of rotation orbit.

In this chapter, a micro-optical gyroscope system with self-calibration is comprised of four PANDA ring resonators and centralized by CC as shown in Figure 7.2. The calibration path design is the core of the calibration technique to design a best test track, in which the errors (misalignments) can be distinguished and measured. In theory, there is a test tracking to filter the best estimate value where the gyroscope calibration uses the angular velocity commands from the gyro output error equation, where a drift is brought by the gyro zero drift, which is irrelevant to the rotation angular velocity. The gyro drift results from the scale factor and nonorthogonal errors,

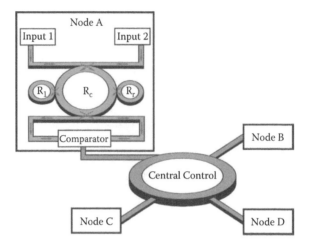

FIGURE 7.2   A schematic of a gyroscope sensor with self-calibration and network control.

which is directly proportional to the angular velocity input. Therefore, we should carefully consider the error source which occurs from the angular velocity, because the large input of angular velocity is difficult for estimating the gyro zero drift, which is submerged in noise, while the small input results in the very long gyro drift.

7.3   RESULTS AND DISCUSSION

From Equation (7.5), there are 12 parameters estimated using the Kalman filter containing the variables that tend to give more precise measurements, in which a series of measurements can be observed over time based on a single measurement alone. The four rotating states of turntable are needed to separate all of these parameters. In general, the four states are chosen as follows:

State 1: The velocity is equivalent to impulse of the gyro zero drift to calibrate the zero drift. The rotation rate and errors are given by $\omega_r = [0\ 0\ 0]^T$ and b_{gx}, b_{gy}, b_{gz}, respectively, where the output result is shown in Figure 7.3.

State 2: The angular velocity rotating around y–axis, in which the rotation rate and errors are given by $\omega_r = [0\ \omega\ 0]^T$ and E_{gxz}, S_{gy}, E_{gzx}, respectively, where the output result is shown in Figure 7.4.

State 3: The angular velocity rotating around x–axis, in which the rotation rate and errors are given by $\omega_r = [\omega\ 0\ 0]^T$ and E_{gyz}, S_{gx}, E_{gzy}, respectively, where the output result is shown in Figure 7.5.

State 4: The angular velocity rotating around z–axis, in which the rotation rate and errors are given by $\omega_r = [0\ 0\ \omega]^T$ and E_{gxy}, S_{gz}, E_{gyx}, respectively, where the output result is shown in Figure 7.6.

In Figure 7.2, each of the gyroscopes is fixed at positions A, B, C, and D, where the information among these gyroscopes is connected by a star network linked by

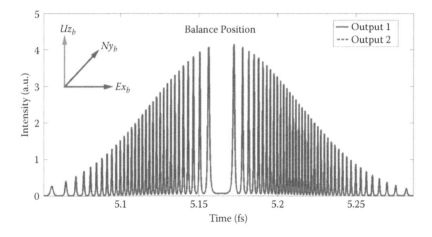

FIGURE 7.3 Result of the steady-state (balance position) signals, where output 1 is the same position in time with output 2.

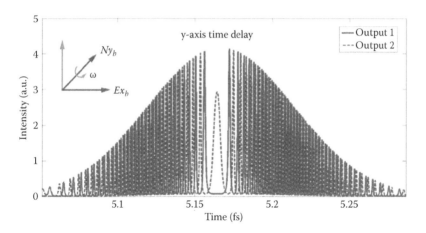

FIGURE 7.4 Result of the rotating signal around y–axis, where output 1 is difference in time with output 2.

optical waveguide (or fiber optic cable). In application, the imbalance positions of each gyroscope can be adjusted by the CC system. First, consider at gyroscope A, the shift in phase (or delay time) can be seen as shown in Figures 7.4, 7.5, and 7.6 for y, x, and z axes, respectively, where similar results can also be obtained at gyroscopes B, C, and D. In this work, the self-calibration technique of the four-point probe gyroscope is performed by using a mathematical model to compensate the gyro error from the misalignment angle (delay time) between the gyro input axis and the initial position. The four-point probe gyroscope is initially in the balance position, which is connected as a star network. When the sensing nodes are rotated, the delay times (phase shift) are as shown in Figures 7.4 through 7.6. The self-calibration

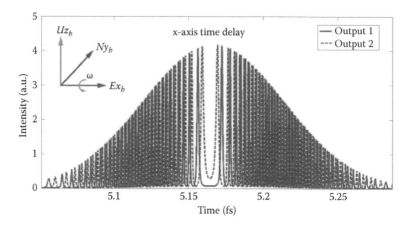

FIGURE 7.5 Result of the rotating signal around x–axis, where output 1 is difference in time with output 2.

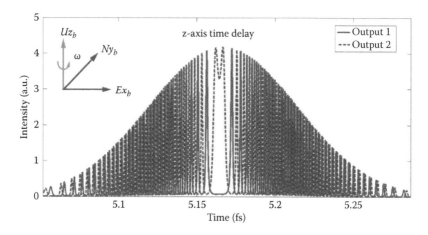

FIGURE 7.6 Result of the rotating signal around z–axis, where output 1 is difference in time with output 2.

of each gyroscope can be performed and backed to the balance position as shown in Figure 7.3. By using the comparator (time delay comparison), the simulation results obtained have shown that the delay times of 2 as, 3 as, and 6 as on y, x, and z axes are seen, respectively. Furthermore, the CC can perform the signal transmission with any node to adjust or control like a network [5], which can be connected through optical waveguide (optical cable) as a system to perform the three-axis adjustment and self-calibration.

In operation, two identical monochromatic light sources are input into the opposite directions of the input ports of the PANDA ring resonator via the input 1 (E_{in1}) and the input 2 (E_{in2}) as shown in Figure 7.1, respectively. At the output 1 (E_{th1})

and the output 2 (E_{th2}), the time differences of those input signals are calculated, which can be explained by the Sagnac effects. First, the gyroscope is considered perfectly still, that is, without movement, in which the two light beams are at the outputs at the same moment. On the other hand, when the gyroscope rotates or tilts, one light signal trip is shortened, that is, shifted or delayed. By measuring the time difference between each return signal, the gyroscope can be determined exactly which way the platform is rolling. By using our proposed gyroscope design, a lot of information about its movement can be determined. Therefore, it is extremely useful for navigation system precision and stabilization. The results of the rotating signal around three axes are as shown in Figures 7.3 through 7.6. Additionally, results of the time and intensity differences of signals from 5.0 to 5.3 fs are shown in Figures 7.7 and 7.8, respectively. Figure 7.7 shows the values of time differences between the starting and ending states, where the median is 2.0 as (attosecond), which means that the high-resolution detection in timescale can be obtained. The median of intensity difference is 0.0546 as shown in Figure 7.8, which is the interval of input and output signals. In our design we have chosen a planar substrate or wafer dimension of 17 μm length and 8 μm width, which consists of two waveguides with 300 nm width and three microring resonators, first and second ring radii with $R_l = R_r = 1.550$ μm, and the third ring radius with $R_c = 3.10$ μm. By using the

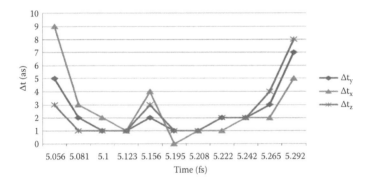

FIGURE 7.7 Graph of the time difference of three axes.

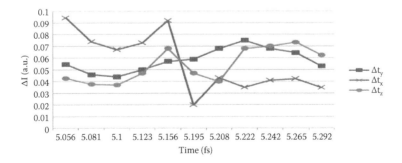

FIGURE 7.8 Graph of the intensity difference of three axes.

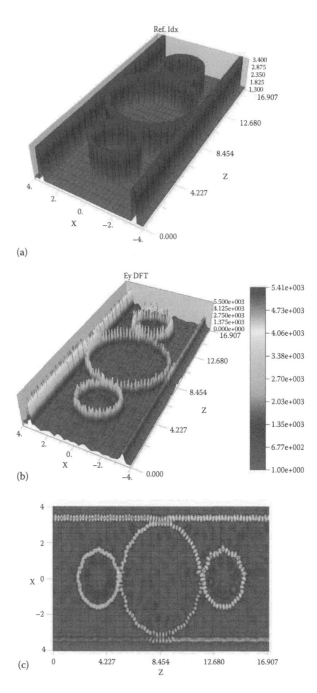

FIGURE 7.9 Optical device testing simulation results in a PANDA ring sensor without any change of cladding refractive index, where (a) the refractive index view of the waveguide is on the wafer, (b) a 3-D image result of DFT, and (c) a 2-D image result of DFT.

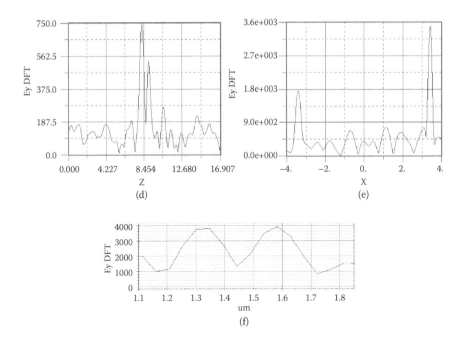

FIGURE 7.9 (Continued) Optical device testing simulation results in a PANDA ring sensor without any change of cladding refractive index, where there is a (d) plane view result along the x and z axes, (e) a plane view result along the Y and Z axes and (f) the field component E_y DFT are at output the cross-section.

Optiwave FDTD applications to design the system layout, we can simulate and analyze the results using the finite-difference time-domain (FDTD) [16]. The input field is an electromagnetic field positioned and oriented to coincide with the input plane, where the input plane is purely a geometrical concept, which is simply a position and a direction, which is fully defined by a plane. The input field is defined by the physical parameters, where the time domain parameters of the input field can be specified as Gaussian modulated continuous wave (GMCW) and a wavelength for the carrier wave by the modal transverse input field.

We consider the case where without any change in cladding refractive index, the optical propagation in the microring sensor (a PANDA) is simulated as shown in Figure 7.9a, which presents the refractive index view of the waveguide on the wafer and uses the input of the simulation. Each waveguide has a material from which the index of refraction is computed based on the material parameters and wavelengths. The obtained results of electric field y component is calculated using discrete Fourier transform (DFT) and shown in Figure 7.9b and c for 3-D and 2-D images, respectively. The plane view result along the x and z, and y and z axes are shown in Figure 7.9d,e. Finally, the field component E_y DFT at the output cross-section is shown in Figure 7.9f. The simulation is performed by using the FDTD method for transverse electrical (TE) wave propagation, in which the input source is a GMCW.

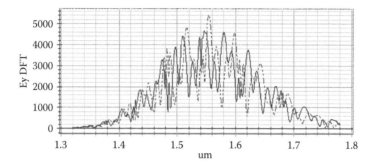

FIGURE 7.10 Shows the relationship between the field component E_y DFT and wavelength of input 1 (E_{in1}) dashed line and input 2 (E_{in2}) solid line.

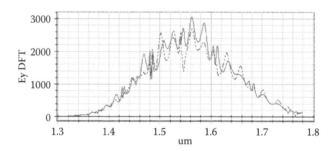

FIGURE 7.11 Shows the relationship between the field component E_y DFT and wavelength of output 1 (E_{th1}) dashed line and output 2 (E_{th2}) solid line.

7.4 CONCLUSION

The new system has been designed for a MOG network, which can be used for four-point probe measurement (see Figures 7.10 and 7.11). By using the four-point probe system, the balancing of four positions (wheels) can be adjusted and controlled dynamically and simultaneously, which is available for vehicle motion control. Simulation results obtained have shown that the phase shift after rotation can be seen and adjusted, in which the calibration method is theoretically feasible, with high accuracy and short time (fs). The advantage is that the fine adjustment can be employed by self-calibration between R_r and R_l signal delay times. According to the proposed system, the use of micro-optical gyroscope with self-calibration control for long-range gyroscope applications, distributed sensors with rotation motion control, and continuous mass motion, especially for vehicle motion, can be realized. In simulation, the physical parameters have been used in practical fabricated devices and realized.

REFERENCES

1. K. Suzuki, K. Takiguchi, and K. Hotate, Monolithically integrated resonator micro-optic gyro on silica planar lightwave circuit, *J. Lightw. Technol.*, 18 (1) 66–72, 2000.

2. H. Liu, L. Feng, Z. Jiao, and R. Li, Polarization noise and reduction technology in micro optical gyroscope, *Nano/Micro Engineered and Molecular Systems*, 956–959, 2011.

3. I. K. Meshkovsky, V. Ye. Strigalev, G. B. Deineka et al., Three-axis fiber-optic gyroscope: Development and test results, *Gyroscopy and Navigation*, 2 (4) 208–213, 2011.

4. N. Song, D. Ma, X. Yi et al., Research on time division multiplexing modulation approach applied in three-axis digital closed-loop fiber optic gyroscope, *Optik*, 121 (23) 2185–2189, 2010.

5. S. Qin, Z. Huang, and X. Wang, Optical angular encoder installation error measurement and calibration by ring laser gyroscope, *IEEE Trans. Instr. and Meas.*, 59 (3) 506–511, 2010.

6. M. Kirkko-Jaakkola, J. Collin, and J. Takala, Bias prediction for MEMS gyroscopes, *IEEE Sensors J.*, 12 (6) 2157–2163, 2012.

7. K. Uomwech, K. Sarapat, and P. P. Yupapin, Dynamic modutated Gaussian pulse propagation within the double PANDA ring resonator, *Microw. Opt. Technol. Lett.*, 52 (8) 1818–1821, 2010.

8. P. Yabosdee, K. Srinuanjan, and P. P. Yupapin, Proposal of the nanosensing device and system using a nano-waveguide transducer for distributed sensors, *Optik*, 121 (23) 2117–2121, 2010.

9. B. Piyatamrong, K. Kulsirirat, W. Techithdeera, S. Mitatha, and P. P. Yupapin, Dynamic potential well generation and control using double resonators incorporating in an add/drop filter, *Modern Phys. Lett. B*, 24 (32) 3071–3082, 2010.

10. T. Phatharaworamet, C. Teeka, R. Jomtarak, S. Mitatha, and P. P. Yupapin, Random binary code generation using dark–bright soliton conversion control within a PANDA ring resonator, *J. Lightw. Technol.*, 28 (19) 2804–2809, 2010.

11. W. P. Lin, M. S. Kao, and S. Chi, The modified star-ring architecture for high-capacity subcarrier multiplexed passive optical networks, *J. Lightw. Technol.*, 19 (1) 32–39, 2001.

12. P. C. Peng, H. Y. Tseng, and S. Chi, A hybrid star-ring architecture for fiber Bragg grating sensor system, *IEEE Photonics Technol. Lett.*, 15 (9) 1270–1272, 2003.

13. J. Scheuer and A. Yariv, Sagnac effect in coupled-resonator slow-light wave guide structures, *Phys. Rev. Lett.*, 96 (5) 05390-1–05390-4, 2006.

14. M. J. F. Digonnet, Rotation sensitivity of gyroscopes based on distributed-coupling loop resonators, *J. Lightw. Technol.*, 29 (20) 3048–3053, 2011.

15. W. Sa-Ngiamsak, C. Sirawattananon, K. Srinuanjan, S. Mitatha, and P. P. Yupapin, Micro-optical gyroscope using a PANDA ring resonator, *IEEE Sensors J.*, 12 (8) 2609–2613, 2012.

16. OptiFDTD by Optiwave Corp.©, ver. 8.0, single license, 2008.

8 Spin Transport Networks

8.1 INTRODUCTION

Electronic spin has been widely investigated and used in many areas of research such as electronic and optical spin coherence manipulations with semiconductors [1], spin transport electronics in metals [2], and electronic spin distributions and magneto-electronic properties [3]. However, the limitation of long-distance links has become a problem for electronic spin; therefore, the search for a new technique remains. In practice, optical spin is recommended as a good candidate, because it has shown an advantage over electronic spin, especially for long-distance transportation without electromagnetic interference in the propagating media. Optical spin is recognized as the promising key for future digital and computing technologies, as it can be used for many applications such as semiconductors [4], magnetic tunnel junctions [5], nano-antenna [6], thin-film nanomagnets [7], and cell communication [8]. Several techniques for optical spin generation have been proposed, where the use of nanoparticle optical excitation with circular light is an interesting idea, in which light pulse is used for spin generation and detection [9]. By using the optical orientation, the angular momentum of light is conversed to electronic spin and vice versa [10], which is very efficient in semiconductors. The consequence of this effect assists an important aspect of spintronics, where it is used to spin-polarize electrons, where the optical spin resonance and transverse spin relaxation in magnetic semiconductor quantum wells is achieved. These techniques were used to study the dynamic spin behavior of photo-injected exactions in the embedded magnetic sublattice [11]. Moreover, the optical spin manipulation in electrically pumped vertical-cavity surface-emitting lasers is analyzed [12], which demonstrated the output polarization mix of electrical and optical excitations. After that, the spin magnetic state and the static magnetic field were used to build the AND, OR, XOR (CNOT), and NAND gates by optical spin manipulation [13].

In a long-haul optical communication network, the current favorite multiplexing technology is wavelength division multiplex (WDM) [14]. All of the end-user equipment needs to operate only at the bit rate of a WDM channel, which can be chosen arbitrarily and at peak processing speed. The new access technologies that can satisfy high performance almost combine WDM technology with high-speed technologies through a hybrid solution, which were reported [15,16]. For quantum information applications, it is essential to use emitters in which the quantum state can be optically controlled. With InAs quantum dot, quantum information can be encoded in the optical spin, which can be completely manipulated using fast optical pulses [17]. The nanophotonic platform consists of devices that provide strong surface plasmon localization of the electromagnetic field [18], connected to an optical network.

Strong localization is essential to enhance the interaction strength between the photons and the metallic coupling. In some applications, metal is coupled to nanoscale optical resonators and can be used as simple optical dipoles that act as efficient light switches or high nonlinear optical media. All of these technologies can be developed as a device with solitons pulse technique for optical communication. The optical spin is proposed for communication systems by Yupapin and his colleagues [19] where the PANDA ring resonator can be used to generate a soliton pair called a *soliton spin* (photonic spin) where Tbit/s high-speed transmission and bandwidth can be achieved when optical amplifiers are combined with WDM in soliton-based communication systems [20].

From previous works, there are several ways to generate optical spin, where most of them are generated by the electron spin using the electronic devices. Recently, Glomglome et al. [21], has proposed that the optical spin can be formed by the orthogonal soliton rotating within the PANDA ring resonator. The optical spin (photonic spin) using a dark–bright soliton pair was realized, where the use of long-distance optical spin networks is possible. In this chapter, the magnetic networks (MagNet) can be formed by optical spins within the optical system, in which the optical spin using a dark–bright soliton pair generates a high magnetic field when the metallic coating is applied on the device surface. A new schematic of magnetic network design is illustrated in the next session. The results illustrated that the soliton spin rotation and the spin sets can be obtained by the proposed system. In application, the magnetic network with soliton spin may be useful for optical fiber communication networks such as nanosensors, nanocommunications, and optical spin cryptography and networks. The theoretical review is also presented with rearranged formalism.

8.2 THEORY

In Figure 8.1, the PANDA ring resonator generates the orthogonal set of dark and bright soliton pairs. The add-drop optical filter device is used to radiate a transverse magnetic field by spin polarization. The output field, E_T and E_1, consists of the transmitted and circulated components within an add-drop optical multiplexing system,

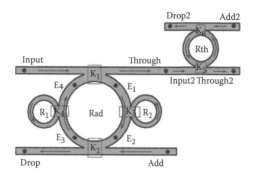

FIGURE 8.1   Schematic of multi-optical spins generated by soliton pulse in a PANDA ring resonator.

which provides the driving force to the photon/molecule/atom. In this study, a modified add-drop filter called a *PANDA* ring resonator which contains a nano- and microring integrated together is proposed. The system achieves a dark–bright soliton conversion process, which produces the orthogonal set of dark and bright soliton pairs and can be decomposed into left and right circularly polarized waves. The relative phase of the two output light signals is $\pi/2$ after coupling into the optical coupler. In addition, the signals produced from the through port and drop port have a π phase change from the original input signal. The concept of orthogonal soliton spins can be assigned as the optical dipoles.

Input and add signals in the form of dark and bright optical solitons are governed by the Equations (8.1) and (8.2), respectively [22], where

$$M = \left[\left(\frac{x}{2L_D}\right) - i\phi(t)\right]$$

in which A and x are the optical field amplitude and propagation distance, respectively. The random phase term related to the temporal coherence function of the input light is as follows,

$$\phi(t) = \phi_0 + \phi_{NL} = \phi_0 + \frac{2\pi n_2 L}{A_{eff}\lambda}|E_0(t)|^2 \tag{8.1}$$

The linear phase shift is ϕ_0, the nonlinear phase shift is ϕ_{NL}, the nonlinear refractive index of InGaAsP/InP waveguide is n_2. The effective mode core area of the device is given by A_{eff}, $L = 2\pi R_{ad}$, R_{ad} is the radius of device, λ is the input wavelength light field, and $E_0(t)$ is the circulated field within nanoring coupled to the right and left add-drop optical filter system as shown in Figure 8.1. T is a soliton pulse propagation time in a frame moving at the group velocity, T_0 is a soliton pulse propagation time at initial input, $T = t - \beta_1 x$, where β_1 and β_2 are the coefficients of the linear and second-order terms of Taylor expansion of the propagation constant. $L_D = T_0^2/|\beta_2|$ is the dispersion length of the soliton pulse, where t is the soliton phase shift time, and the frequency shift of the soliton is ω_0. This solution describes a pulse that keeps its temporal width invariance as it propagates, and thus it is called a *temporal soliton*. The refractive index (n) of light propagates within the medium given by

$$n = n_0 + n_2 I = n_0 + \frac{n_2}{A_{eff}}P \tag{8.2}$$

where n_0 and n_2 are the linear and nonlinear refractive indexes. I and P are the optical intensity and optical power, respectively. A_{eff} is the effective mode core area of the system. For the microring resonator (MRR) and nanoring resonator (NRR), the effective mode core areas are ranged from 0.50 to 0.10 μm^2 [23,24].

Analytically, for the light pulse passes through the coupler of the PANDA multiplexing system [25], the transmitted and circulated components can be written as

$$E_T = Y_1\left[X_1 E_{in1} + \sqrt{\kappa_1}E_4\right] \tag{8.3}$$

$$E_1 = Y_1 \left[X_1 E_4 + \sqrt{\kappa_1} \, E_{in1} \right] \tag{8.4}$$

$$E_2 = E_{R2} E_1 \exp\left[-\frac{a}{2}\frac{L}{2} - j\kappa_n \frac{L}{2} \right] \tag{8.5}$$

where

$$Y_1 = \sqrt{1-\gamma_1}, \, X_1 = \sqrt{1-\kappa_1}$$

For the second coupler of the add-drop optical multiplexing system

$$E_D = Y_2 \left[X_2 E_{i2} + \sqrt{\kappa_2} \, E_2 \right] \tag{8.6}$$

$$E_3 = Y_2 \left[X_2 E_2 + \sqrt{\kappa_2} \, E_{i2} \right] \tag{8.7}$$

$$E_4 = E_{R1} E_3 \exp-\frac{\alpha}{2}\frac{L}{2} - j\kappa_n \frac{L}{2} \tag{8.8}$$

where

$$Y_2 = \sqrt{1-\gamma_2}, \, X_2 = \sqrt{1-\kappa_2}$$

where κ_1 and κ_2 are the intensity coupling coefficient, γ_1 and γ_2 are the fractional coupler intensity loss, α is the attenuation coefficient, $\kappa_n = 2\pi/\lambda$ is the wave propagation number, λ is the input wavelength light field, and $L = 2\pi R_{ad}$, which R_{ad} is the radius of add-drop device. The circulated light fields, E_{R1} and E_{R2}, are the light field circulated components of the nanoring radii, R_1 and R_2, which are coupled into the left and right sides of the add-drop optical multiplexing system, respectively. The light field transmitted and circulated components in the right nanoring, E_{R1} and E_{R2}, are given by

$$E_{R1} = E_3 \left\{ \frac{\left[\sqrt{Y_4 X_4} - Y_4 \right] \exp - \left(\frac{a}{2} L_1 - j\kappa_n L_1 \right)}{\left[1 - \sqrt{Y_4 X_4} \right] \exp - \left(\frac{a}{2} L_1 - j\kappa_n L_1 \right)} \right\} \tag{8.9}$$

$$E_{R2} = E_1 \left\{ \frac{\left[\sqrt{Y_3 X_3} - Y_3 \right] \exp - \left(\frac{a}{2} L_2 - j\kappa_n L_2 \right)}{\left[1 - \sqrt{Y_3 X_3} \right] \exp - \left(\frac{a}{2} L_2 - j\kappa_n L_2 \right)} \right\} \tag{8.10}$$

where $Y_3 = (1 - \gamma_3)$, $Y_4 = (1 - \gamma_4)$, $X_3 = (1 - \kappa_3)$, κ_3 and κ_4 are the intensity coupling coefficient, γ_3 and γ_4 are the fractional coupler intensity loss, α is the attenuation coefficient, $\kappa_n = 2\pi/\lambda$ is the wave propagation number, λ is the input wavelength light field, and $L_1 = 2\pi R_1$, R_1 is the radius of the left nanoring.

The output optical field (E_T) and power output (P_T) of the through port are expressed by

$$E_T = x_1 y_1 E_{in1} + (jx_1 x_2 y_2 \sqrt{\kappa_1} E_{R2} E_{R1} E_1 - x_1 x_2 \sqrt{\kappa_1 \kappa_2} E_{R1} E_{add}) \exp(TP) \tag{8.11}$$

$$P_T = (E_T) \cdot (E_T)^* = \left| x_1 y_1 E_{in1} + \left(jx_1 x_2 y_2 \sqrt{\kappa_1} E_{R2} E_{R1} E_1 \right. \right.$$

$$\left. \left. -x_1 x_2 \sqrt{\kappa_1 \kappa_2} E_{R1} E_{add} \right) \exp(TP) \right|^2 \tag{8.12}$$

The output optical field (E_D) and power output of the drop port (P_D) are given by:

$$E_D = x_2 y_2 E_{add} + jx_2 \sqrt{\kappa_2} E_{R2} E_1 \exp(TP) \tag{8.13}$$

$$P_D = (E_D) \cdot (E_D)^* = \left| x_2 y_2 E_{add} + jx_2 \sqrt{\kappa_2} E_{R2} E_1 \exp(TP) \right|^2 \tag{8.14}$$

where

$$TP = -\frac{a}{4}\frac{L}{4} - j\kappa_n \frac{L}{4}$$

In order to retrieve the required signals, the proposed system used an add-drop optical multiplexing device, in which the electric field (Et_2) and light pulse output power (Pt_2) are given by:

$$E_{t2} = E_{t1} \frac{-X_5 \exp\left[-\frac{\alpha}{2} L_b - j\kappa_n L_b \right] + X_5}{1 - X_5 X_6 \exp\left[-\frac{\alpha}{2} L_b - j\kappa_n L_b \right]} \tag{8.15}$$

$$P_{t2} = \left| \frac{E_{t2}}{E_{t1}} \right|^2 = \frac{(1 - \kappa_5 - 2X_5 X_6 \exp PX \cos(\kappa_n L_b) + (1 - k_5) \exp P)}{(1 + (1 - \kappa_5)(1 - \kappa_6) \exp P - 2X_5 X_6 \exp P \cos(\kappa_n L_b))} \tag{8.16}$$

The electric field detected (Ed_2) and the light pulse output power (Pd_2) are expressed by:

$$E_{d2} = E_{t1} \frac{-\sqrt{\kappa_5 \kappa_6} \exp\left[-\frac{\alpha}{2}\frac{L_b}{2} - j\kappa_n \frac{L_b}{2} \right]}{1 - X_5 X_6 \exp\left[-\frac{\alpha}{2} L_b - j\kappa_n L_b \right]} \tag{8.17}$$

$$P_{d2} = \left| \frac{E_{d2}}{E_{t1}} \right|^2 = \frac{\kappa_5 \kappa_6 \exp P}{(1 + (1 - \kappa_5)(1 - \kappa_6) \exp P - 2X_5 X_6 \exp P \cos(\kappa_n L_b))} \tag{8.18}$$

given

$$X_5 = \sqrt{1 - \kappa_5}, \ X_6 = \sqrt{1 - \kappa_6}, \ \text{and} \ P = \left[-\frac{\alpha}{2} L_b \right]$$

where $L_b = 2\pi R_b$, R_b is the radius of the add-drop optical multiplexing as shown in Figure 8.1.

8.3   SPIN TRANSPORT NETWORKS

In simulation, the parameters for the add-drop optical multiplexer and both nanorings on the left- and right-hand sides of the PANDA ring are set at $R_1 = R_2 = 2.5$ μm, $R_{th} = 15$ μm, and radius of the center ring is $R_{ad} = 30$ μm, respectively. Coupling coefficient ratios are $\kappa_1 = \kappa_4 = 0.3$, $\kappa_2 = \kappa_3 = 0.5$, effective core area of the waveguides is

$A_{eff} = 0.25 \ \mu m^2$, the refractive index value of InGaAsP/InP is 3.14 [26,27]. Waveguide loss coefficient $\alpha = 0.1$ dB/mm and center wavelength is chosen to be at $\lambda_0 = 1400$, 1425, 1450, 1475, 1500, 1525, 1550, 1575, and 1600 nm. A dark soliton light pulse with 10 mW peak power is input into the input port, travels and passes through the first coupler, κ_1, where it splits into two parts; one part goes through the port and the other is to the arc at E_1 of the add-drop optical multiplexer. The light pulse enters and circulates in the right nanoring where it passes to the add-drop optical multiplexer at R_r. The output power is amplified (larger amplitude). To use the control function, a bright soliton with 1 W peak power is input into the control port, passes through the second coupler, κ_2, and it is multiplexed with the light pulse from E_2. The output light is then split into two parts. One part goes to the drop port, E_d, the power output, and the other is to the arc of the add-drop at position E_3. After that, the light pulse travels and enters into the left nanoring resonator radii, R_l, and then passes to the add-drop optical multiplexer at position E_4. The output power is amplified again and travels into the first coupler, κ_1. It then enters into the through port, E_t and E_1. It has been observed that when the dark soliton arrays are fed into a PANDA ring, the bright soliton pulses are detected at E_1 and E_4 and the dark soliton pulses are detected at E_2 and E_3 position as shown in Figure 8.2. In Figure 8.3, a set of spins is obtained at through (spin-up) and drop (spin-down) ports, respectively. From this result, we managed to show that many soliton spins can be produced from the PANDA ring resonator, where the output signals are randomly collected at both the throughput and drop port of the system. Thus, transverse electric (TE) and transverse magnetic (TM) fields of the solitons corresponding to the left-hand and right-hand photons are produced. Whenever a photon is absorbed by an object, there are two possible optoelectronic spin states that will be exhibited and imparted to the object which correspond to the

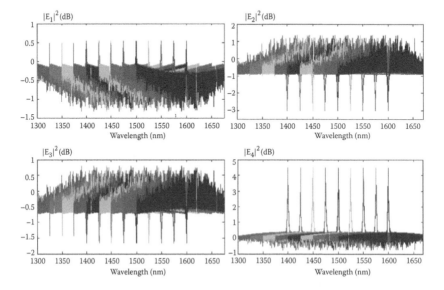

FIGURE 8.2 Many soliton spins within a PANDA ring are generated using a dark-soliton pump input at center wavelength 1400, 1425, 1450, 1475, 1500, 1525, 1550, 1575, and 1600 nm.

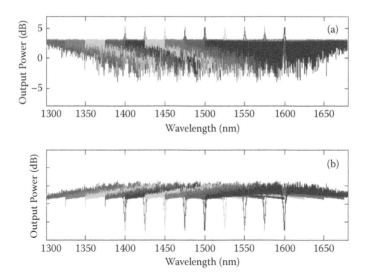

FIGURE 8.3 A set of spins obtained at (a) through port (spin up) and (b) drop ports (spin down).

+ℏ or −ℏ angular momentum. The angular momentum of either +ℏ or −ℏ is imparted to the object when a photon is absorbed by an object. Hence, the array of soliton spins can be generated and controlled by the proposed system, which is available for high-density spin states use.

Many spin results are obtained by using the proposed add-drop optical filter device as shown in Figure 8.4. A realistic current density of an antenna is obtained by using the Pocklington's integral equation [28]. The calculation of the complex one-dimensional current density is performed by using the MATLAB function [29]. The antenna input impedance is defined by the simplify impedance of the antenna for transmission at different lengths [30]. The optical intensity from the PANDA ring resonator is fed into the nano-antenna. The most popular material used for nano-antenna is gold, which gives a good effective resonance and dielectric constant values. The Drude model in this chapter is used to explain the data [31], while the dispersion relation of the surface plasmon resonance at the interface is given by

$$\varepsilon_m(\omega) = \varepsilon_1 + i\varepsilon_2 \tag{8.19}$$

$$\varepsilon_1(\omega) = 1 - \frac{\omega_P^2}{\omega^2 + \Gamma^2} \tag{8.20}$$

$$\varepsilon_2(\omega) = \frac{\Gamma\omega_P^2}{\omega\left(\omega^2 + \Gamma^2\right)} \tag{8.21}$$

where ω_p and Γ denote the size and temperature-dependent plasma and collision frequency (f) of the nanoparticles, that is, $\omega_p = 1.36 \times 10^{16}$ rad s^{-1}, $\omega_p = 2\pi f$, and $\Gamma = 1.05 \times 10^{14}$ rad s^{-1}.

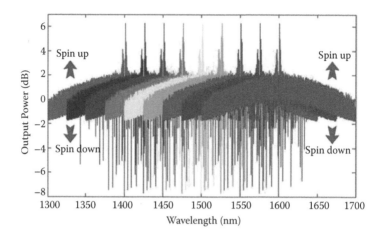

FIGURE 8.4   A set of spins obtained at through and drop ports of the Rth add-drop.

The surface plasmon resonance intensity enhanced absorption of gold nano-antenna increases the photothermal conversion process efficiency. The extinction cross-section of metal nanoparticle is derived by Mie theory given by

$$E(\omega) = \frac{24\pi R \varepsilon_m^{3/2}}{\lambda} \frac{\varepsilon_2(\omega)}{\left(\varepsilon_2(\omega) + 2\varepsilon_m\right)^2 + \varepsilon_2^2(\omega)} \tag{8.22}$$

A metal-dielectric interface supports charge density oscillations along the interface in which the quantum is called *surface plasma oscillations*. The surface plasmons are accompanied by a longitudinal electric field which decays exponentially in metal as well as in dielectric medium. The electric field has its maximum value at metal dielectric interface. The TM-polarization and exponential decay of the electric field are calculated by solving the Maxwell equation for semi-infinite media of metal and dielectric with an interface of metal dielectric. The propagation constant (Kp) of the surface plasmon wave propagating along the metal-dielectric interface is given by

$$\kappa_p = \kappa_0 \sqrt{\left(\frac{\varepsilon_d \varepsilon_m(\omega)}{\varepsilon_d + \varepsilon_m(\omega)}\right)} \tag{8.23}$$

where κ_0 is the wave vector in air, ε_d is the relative permittivity of the dielectric, and $\varepsilon_m(\omega)$ is the dispersive relative permittivity of the metal. This wave can be optically excited using a coupling coated with a thin gold film (Kretschmann configuration), whereas a TM-polarized light beam is imposed on the gold face under total internal reflection conditions. Figure 8.5 shows the 3-D drawing of a ring laterally gold coupled to the straight waveguides (Figure 8.5a) and cross-section of add-drop (Figure 8.5b). Initially, the ring was coated with gold on the top, in which the output spin states can be confirmed by using the output polarizing beam splitter arrangement.

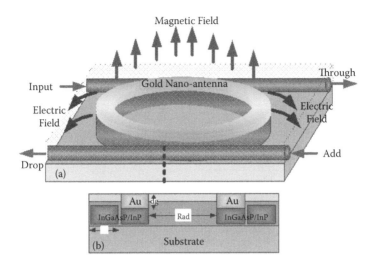

FIGURE 8.5 3-D drawing of ring laterally gold coupled to the straight waveguides (a) and cross-section of add-drop (b).

The orthogonal soliton pair known as a dark–bright soliton conversion pulse within a PANDA ring resonator is generated. The increase in optical signals can be obtained using the nonlinear ring devices beside the conventional add-drop filter, where many optical spins can be generated and propagated into the provided network. The add-drop filters are employed to retrieve the required signals, where the different wavelength signals (spins) can be obtained using the wavelength selectors (add-drop filters) at the end of the drop port. Alternatively, the modified data can be input into the network via the add port. Figures 8.6 and 8.7 show the results that were received by each user 1–4 from the add-drop filter at the through and drop ports in the transmission networks as shown in Figure 8.8, where the spin wavelength selectors are employed as the key instrumental components, in which certain wavelengths or a narrow band of wavelengths can be obtained before the required spins are detected via the end users. For instance, by using an optical filter, that is, polarizing beam splitter, the simulated transmission FSR = 25 nm (free spectral range) and FWHM = 4 nm (full width at half maximum) can be obtained as shown in Figure 8.9, where the output optical spins with different wavelengths can be obtained.

In operation, the optical spins and the magnetic fields can be generated by applying a gold film coating, which induces the magnetic induction on the device surface as shown in Figure 8.5. In Figure 8.8, a polarizing beam splitter (PBS) is used to distinguish the data and referencing signals to the destination (Figure 8.8a). Later, data from the device enters into the multiplexing device (multiplexer, MUX) to encrypt data and transmit into the optical network. The generated multispins (many optical spins) can be used to form the wavelength division multiplexing (WDM) data transmission, which offers the advantage of a long journey without interference and safely arrives at the destinations. Finally, the required signals can be retrieved (decoded

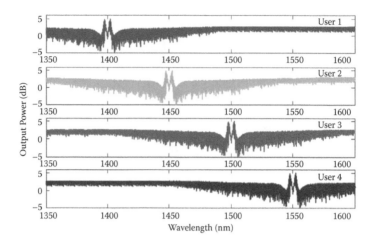

FIGURE 8.6   This shows the results received by each user 1–4 from the add-drop filter at the through port.

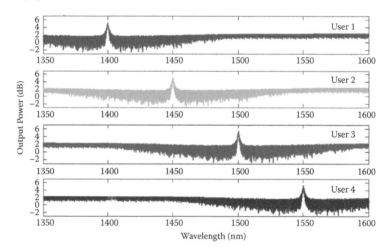

FIGURE 8.7   This shows the results received by each user 1–4 from the add-drop filter at the drop port.

by the demultiplexing, DEMUX) in part b (Figure 8.8b). After traveling through the DEMUX, the information will also be sent to the end users using the add-drop filters, in which the magnetic field is generated to identify the required data in part c (Figure 8.8c). In applications, the new scheme of magnetic resonance imaging (MRI) devices can be formed using the magnetic networks (MagNets), where we can use the magnetic field subrogate to form the large magnetic networks. The new design MRI devices in the future will be small, portable, low cost, and high performance.

To provide a sufficient electric field drop between the optical filter and effective detector, therefore, the same parameters with the simulation results from Mikroulis, Roditi, and Syvridis [27] were used in this work. Generally, the long-distance link

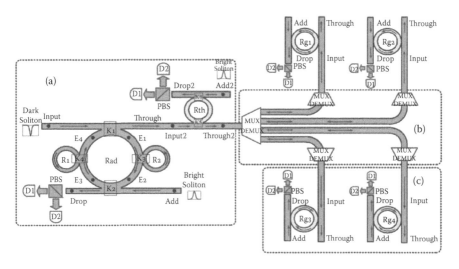

FIGURE 8.8 A schematic diagram of spin distributed networks using the multi-optical spins, where Di: detectors, PBS: polarizing beam splitter, Rgs: output ring radii (users).

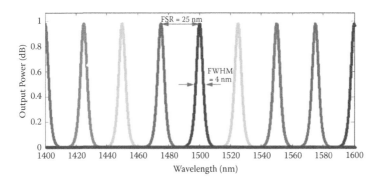

FIGURE 8.9 Simulated transmission FSR = 25 nm and FWHM = 4 nm.

means the soliton output power can be useful for >1000 km via an optical fiber link [32], where the proposed system power of 10 dBm is obtained, which can be transmitted over 4000 km for nine-channel transmission [33], while the time jitter in a soliton communication can be reduced by using the periodic optical phase conjugation (OPC) device [21]. By using the small waveguide effective area, the polarization-dependent loss of <0.1 dB at 1550 nm can be achieved [23,34], where the spin detection can be realized in the form of a polarization mode by using the polarizing beam splitter and can be detected by the end users. The minimum detuning of the two signals (solitons) can be obtained, where in this case the detuning of 6.8 THz can be obtained between two signals (bands), where more details can be found in Yang et al. [35]. However, this is a simulation work in practice, if the propagation loss varies with wavelength, then the solitons away from the center (peak) wavelength (1500 nm) will not be valid for long-distance propagation.

8.4 CONCLUSION

In this work, a magnetic network (MagNet) design system is proposed using the orthogonal dark–bright soliton conversion pulses within a PANDA ring resonator. The magnetic fields are induced by applying metallic coating on the surface of the device. The coupling between the optical fields and the metal coating waveguide generates the transverse electric (TE) and magnetic field (TM) that can be separated using the add-drop filter and PBS for reference source and required signals at the destinations (end users). The results obtained have shown that multi-optical spins in the optical networks, that is, spin networks (SpinNet), can be formed. In addition, the MagNet with soliton spin may be beneficial for optical network communication and cryptography-based spins. The advantage of the optical spin in network is that there is no electromagnetic interference during propagation within the network, which is appropriate for long-distance links.

REFERENCES

1. V. A. Sih, E. Johnston-Halperin, D. D. Awschalom, Optical and electronic manipulation of spin coherence in semiconductors, *Proceedings of the IEEE* 91, 752–760, 2003.
2. M. Johnson, Overview of spin transport electronics in metals, *Proceedings of the IEEE,* 91, 652–660, 2003.
3. L. A. Pozhar and W. C. Mitchel, Collectivization of electronic spin distributions and magneto-electronic properties of atomic clusters of Ga and In with As, V, and Mn, *IEEE Transactions on Magnetics,* 43, 3037–3039, 2007.
4. R. C. Myers, M. H. Mikkelsen, J. M. Tang, A. C. Gossard, M. E. Flatte, and D. D. Awschalom, Zero-field optical manipulation of magnetic ions in semiconductors, *Nat. Mater.,* 7, 203–208, 2008.
5. D. F. Gordon, B. Hafizi, and A. Ting, Nonlinear conversion of photon spin to photon orbital angular momentum, *Opt. Lett.,* 34, 3280–3282, 2009.
6. N. Thammawongsa, N. Moongfangklang, S. Mitatha, and P. P. Yupapin, Novel nano-antenna system design using photonic spin in a PANDA ring, *PIER Lett.,* 31, 75–87, 2012.
7. O. Ozatay, P. G. Gowtham, K. W. Tan, J. C. Read, K. A. Mkhoyan, M. G. Thomas, G. D. Fuchs, P. M. Braganca, E. M. Ryan, K. V. Thadan, J. Silcox, D. C. Ralph, and R. A. Buhrman, Sidewall oxide effects on spin-torque and magnetic-field-induced reversal characteristics of thin-film nanomagnets, *Nat. Mater.,* 7, 567–573, 2008.
8. K. Meyl, Task of the introns, cell communication explained by field physics, *J. Cell Commun. Signal.,* 6, 53–58, 2012.
9. G. Lampel, Nuclear dynamic polarization by optical electronic saturation and optical pumping in semiconductors, *Phys. Rev. Lett.* 20, 491–493, 1986.
10. E. J. Galvez and N. Zhelev, Orbital angular momentum of light in optics instruction, *J. Opt. Soc. Am. B: Opt. Phys.,* 1, 1–6, 2007.
11. S. A. Crooker and D. D. Awschalom, Optical spin resonance and transverse spin relaxation in magnetic semiconductor quantum wells, *Phys. Rev. B: Condens. Matter.,* 56, 7574–7587, 1997.
12. S. Hövel, A. Bischoff, N. C. Gerhardt, M. R. Hofmann, T. Ackemann, Optical spin manipulation of electrically pumped vertical-cavity surface emitting Lasers, *Appl. Phys. Lett.,* 92, 041118–3, 2008.
13. W. Hübner, S. Kersten, and G. Lefkidis, Optical spin manipulation for minimal magnetic logic operations in metallic three-center magnetic clusters, *Phys. Rev. B: Condens. Matter.,* 79, 184431–5, 2009.

14. F. Xiong, W. D. Zhong, and H. Kim, A broadcast-capable WDM passive optical network using offset polarization multiplexing, *J. Lightwave Technol.*, 30, 2329–2336, 2012.

15. E. Wong, Next-Generation Broadband Access Networks and Technologies, *J. Lightwave Technol.*, 30, 597–608, 2012.

16. Z. A. El-Sahn, J. M. Buset, and David V. Plant, Overlapped-subcarrier multiplexing for WDM passive optical networks: Experimental verification and mathematical analysis, *J. Lightwave Technol.*, 30, 754–763, 2012.

17. L. Hai-Feng, K. Sha-Sha, Z. Xiao-Tao, and Z. Huai-Wu, Spin-polarized transport through a two-level quantum dot driven by ac fields, *J. Appl. Phys.*, 109, 054305–6, 2011.

18. G. Armelles, J. B. González-Díaz, A. García-Martín, J. M. García-Martín, A. Cebollada, M. Ujué González, S. Acimovic, J. Cesario, R. Quidant, and G. Badenes, Localized surface plasmon resonance effects on the magneto-optical activity of continuous Au/Co/Au trilayers, *J. Opt. Soc. Am. B: Opt. Phys.*, 1, 16104–12, 2008.

19. F. D. Muhammad, C. Teeka, J. Ali, and P. P. Yupapin, Optical spin manipulation using dark–bright soliton behaviors in a panda ring resonator, *Microwave Opt. Technol. Lett.*, 54, 987–990, 2012.

20. I. Morita, M. Suzuki, N. Edagawa, K. Tanaka, and S. Yamamoto, Long-haul soliton WDM transmission with periodic dispersion compensation and dispersion slope compensation, *J. Lightwave Technol.*, 17, 80–85, 1999.

21. S. Glomglome, I. Srithanachai, C. Teeka, S. Mitatha, S. Niemcharoen, and P. P. Yupapin, Optical spin generated by a soliton pulse in an add–drop filter for optoelectronic and spintronic, *Opt. Laser Technol.*, 44, 1294–1297, 2012.

22. K. Sarapat, N. Sangwara, K. Srinuanjan, P. P. Yupapin, and N. Pornsuwancharoen, Novel dark–bright optical solitons conversion system and power amplification, *Opt. Eng.*, 48, 045004–7, 2009.

23. Y. Kokubun, Y. Hatakeyama, M. Ogata, S. Suzuki, and N. Zaizen, Fabrication technologies for vertically coupled microring resonator with multilevel crossing busline and ultracompact ring radius, *IEEE J. Quantum Electron.*, 11, 4–10, 2005.

24. C. Fietz and G. Shvets, Nonlinear polarization conversion using microring resonators, *Opt. Lett.*, 32, 1683–1685, 2007.

25. T. Phatharaworamet, C. Teeka, R. Jomtarak, S. Mitatha, and P. P. Yupapin, Random binary code generation using dark–bright soliton conversion control within a PANDA ring resonator, *J. Lightwave Technol.*, 28, 2804–2809, 2010.

26. J. Zhu, S. K. Ozdemir, Y. F. Xiao, L. Li, L. He, D. R. Chen, and L. Yang, On-chip single nanoparticle detection and sizing by mode splitting in an ultrahigh-Q microresonator, *Nat. Photonics.*, 32, 1683–1685, 2010.

27. S. Mikroulis, E. Roditi, and D. Syvridis, Direct modulation properties of 1.55µm InGaAsP/InP microring lasers, *J. Lightwave Technol.*, 26, 251–256, 2008.

28. H. C. Pocklington, Electrical oscillations in wires, *Proc. Cambridge Philos. Soc.*, 9, 324–332, 1897.

29. P. Biagioni, J. S. Huang, and B. Hecht, Nano antennas for visible and infrared radiation, *Rep. on Prog. Phys.*, 75, 24402–24441, 2012.

30. C. A. Balanis, *Antenna theory, Rev. Process. IEEE.*, 80, 7–22, 1992.

31. A. D. Rakic, A. B. Djuriic, and J. M. Elazar, Optical properties of metallic films for vertical-cavity optoelectronic devices, *Appl. Opt.*, 37, 5271–5283, 1998.

32. V. S. Grigoryan, R. M. Mu, G. M. Carter, and C. R. Menyuk, Experimental demonstration of long-distance dispersion-managed soliton propagation at zero average dispersion, *IEEE Photon. Technol. Lett.*, 12, 45–46, 2000.

33. R. J. Essiambre and G. P. Agrawal, Timing jitter of ultrashort solitons in high-speed communication systems. II. Control of jitter by periodic optical phase conjugation, *J. Opt. Soc. Am. B*, 14, 323–320, 1997.

34. F. Van Laere, T. Stomeo, C. Cambournac, M. Ayre, R. Brenot, H. Benisty, G. Roelkens, T. F. Krauss, D. Van Thourhout, and R. Baets, Nanophotonic polarization diversity demultiplexer chip, *IEEE Photon. Technol.*, 27, 417–425, 2009.
35. W. X. Yang, J. M. Hou, Y. Y. Lin, and R. K. Lee, Detuning management of optical solitons in coupled quantum wells, *Phys. Rev. A.*, 79, 033825, 2009.

9 Molecular Motor for Drug Delivery

9.1 INTRODUCTION

A molecular motor is recognized as an essential agent of living organ movement, especially for cell communication and signaling. Commonly, the cell structure called *cytoskeleton* has a function to serve the track of network in intracellular transport processes and it plays a crucial role in motility of the cell, which works as a machine [1–2]. The cytoskeleton within the axon and dendrite cooperating with a motor protein can move along the substrate including the actin filament (7 to 9 nm in diameter), microtubule (25 nm in diameter), and intermediate filament (10 nm in diameter) [3]. There are three significant super families of cytoskeleton molecular motor protein, which are myosin, kinesin, and dynein [4]. Molecular motor manipulation has been extensively investigated by many research techniques, which contributed many advances to the study of the biophysical properties of molecular motors [5–7]. In general, the manipulation of molecular motors can be characterized by five elements: first, the energy input type supplied to work [8]; second, the type of motion performed by the used components [9]; third, the monitoring methods used for the operation [10]; then, there is the possibility of the repeated operation in cycles [11], and last, the average desired time scale to complete a cycle [12]. Furthermore, there is much interesting research showing that the motor movement direction [13–14] can be controlled, which has been quantitatively analyzed by recording the relative length changes of DNA using laser tweezers or magnetic devices, which allows the individual consecutive chemical and mechanical steps of the motor enzymes to be dissected.

The microscopic manipulation concept was first proposed by Ashkin in 1987, in which he presented the optical trap that became the "optical tweezers" for manipulating biological objects [15], where it was shown that bacteria and viruses were trapped using an argon laser at a wavelength of 514 nm. However, the visible laser caused substantial damage to the biological objects even at very low powers. After that, the new era of molecular motor control by light was established, where several studies of different dynamics and biomolecular processes, ranging from how individual macromolecules such as proteins, DNA, and RNA unfold under force [11, 16] was realized, and in which it was shown how molecular motors translocate and exert forces [17]. The particular interest was that the single-molecule optical trap experiment provided the novel insights into the mechanism of nucleic acid translocation and related research for trapping nanoparticles [18]. Magnetic tweezers are the interesting technique to implement for reducing the optical damage induced in a trapped particle [19]. The magnetic tweezers consist of a pair of permanent magnets placed above the sample holder, which inverts the microscope outfitted with a charge-coupled device. Magnetic tweezers are capable of forming the applied forces of 1 nN, which can be used to manipulate and rotate magnetic particles ranging from 0.5 to 5 nm. In addition, the tweezer technique can be used *in vivo* for a living organ and fabricated in a nanoscale regime. The Laguerre–Gaussian beam technique [20] is also available, where all photons have intrinsic angular momentum called *spin*, so a circularly polarized laser beam can also have spin angular momentum. On the other hand, a laser beam with orbital angular momentum can possibly be formed, where the involved torque is strong enough to make microscopic particle rotation.

Recently, the modified add-drop optical filter known as a PANDA ring resonator was proposed by Jalil et al. [21], which is capable of generating dynamic optical tweezers (potential wells) in order to trap the nanoparticles. Subsequently, a new optical trapping design to transport gold nanoparticles using a PANDA ring resonator system was reported by Aziz et al. [22], where intense optical fields in the form of dark solitons can be controlled by Gaussian pulses used to trap and transport nanoscopic volumes of matter to the desired destination via an optical waveguide. Although there is much in the literature about molecular motor manipulations, rarely is there research investigating molecular motor controlled by light. In this chapter, we have demonstrated that the dynamic behavior of the tweezers can be rotated in the same ways as the photonic (optical) spin [23], which can be available for long-distance molecule trapping and transportation along the optical waveguide, especially for drug delivery, diagnosis, and therapeutic applications [24–26]. Finally, a new concept of molecular motor control using dynamic optical tweezers within a modified optical add–drop filter called a *PANDA ring resonator* is established; the optical tweezers are generated by dark and bright soliton pairs corresponding to the left-hand and right-hand rotating solitons. The carrier signals in the form of optical vortices or potential wells can be used to trap and rotate molecules by the plasmonic surfaces and transport them along the optical waveguide. The merit of such a system is its highly stable signal with no fluctuation over a certain period of time, where the trapped atom or molecule is confined during the delivery process. In application, the trapped molecules can move and rotate to the required destinations, which can be useful for many applications, especially for drug delivery, medical diagnosis, and therapy.

9.2 MOLECULAR MOTOR

Generally, myosin moves on the actin filaments, while both kinesin and dynein move on the microtubules. Most of these molecular motors are dimers with two *heads* connected together at a *stalk* region and a *tail* domain opposite the heads. The head of the molecular motor contains the motor domain that provides the motion along the filaments whereas the tail of the molecular motor contains the subunits responsible for cargo binding and regulation. Molecular motors can take hundreds of steps along the respective filament in long-distance cellular cargo transport before detaching. Some cargos such as vesicles, mitochondria, mRNA, virus particles, and endosomes [27] are moved by such motors. The movement of kinesins is in a hand-over-hand mechanism, which is interesting to understand more, see Nobutaka et al. [12]. Dynein molecular motors are divided into two groups: (i) cytoplasmic dyneins, which achieve various intracellular cargo transport functions; and (ii) axonemal dyneins, which power the motion of cilia and flagella in some eukaroytic cells and are attached in large linear arrays along the microtubule inside the cilia and flagella. All dyneins walk toward the minus-end of the microtubule. As shown by electron microscopy, it utilizes the reconstructions of cytoplasmic dynein, which demonstrates a structure similar to axonemal dynein. Hence, the inducement of mechanical torque of their functions occurs in a similar way [28]. The myosin molecular motors function in a variety of cellular tasks, from cellular transport to muscle contraction. Consequently, the myosins are divided into two groups: (i) nonmuscle myosins, which are involved in organelle transport along actin filaments very similar to the mechanism of kinesins; and (ii) muscle type myosins that drive muscle contractions and are an important component of the muscle. These cellular movements, cellular transport, and muscle contractions depend on the interactions between actin filaments and myosin [29]. The transportation process is vital for the cell because it confines selected organelles with appropriate spatial-temporal coordinates. Currently, a variety of research on motor performance focuses on the mechanochemical cycle to perform mechanical movements under the control of appropriate energy inputs [30] and for the duration of the motor protein along the substrate, which on average is forward moving [31].

Figure 9.1 shows the axonal transport of membranous *in vivo*, where the cross-bridge structures between microtubules and membranous organdies have been identified by the electron microscopic technique which is a candidate for membranous organelle translocations. At present, kinesin and dynein are well known in neurons as microtubule motors that produce the force necessary for the fast axonal transport of membranous organelles. Dyneins can move toward the minus end of the microtubule, where they transport cargo from the cell margin toward the center from the neuronal axon terminal to the cell body (soma), which is known as *retrograde transport*. A different type of motor protein known as *kinesin* moves to the plus end

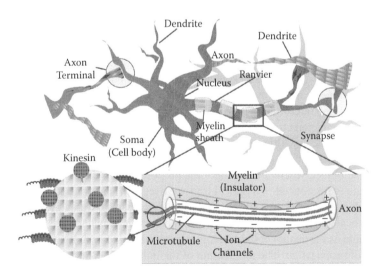

FIGURE 9.1   Axonal transport model of membranous organelle for cell communications.

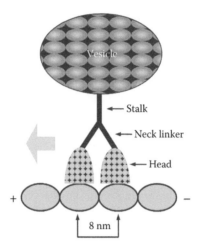

FIGURE 9.2 Schematic of gross anatomy of a kinesin motor protein.

of microtubules, which is associated with the anterogradely moving membranous organelles in the axon, so it is considered to be an anterograde translocation. The microtubule structure is seen in the polar form and the head is only connected to the microtubule in one orientation, while the ATP binding gives each step of its direction through a process called *neck linker zippering* as shown in Figure 9.2.

The conventional kinesin and other members of the kinesin family bind ATP and microtubules at a specific site in their conserved motor domain, and use the energy from ATP hydrolysis to produce force and to move along the microtubules. There are 45 different kinesin species in humans, where they can be distributed in all eukaryotic cells by linear motors, which are involved in many functions of biological systems, including cargo transport, microtubule dynamics control, and mitosis, and they play a crucial role in ways of signal communication. Kinesins have a wide variety of structures, and they function as monomers, dimers, or tetramers in cells, which are classified into three categories: N-terminal kinesins, C-terminal kinesins, and M-kinesins. N-terminal kinesins move to the plus end of the microtubule, and C-terminal kinesins move toward the minus end of the microtubule [12]. When the kinesin motor and microtubule track have interacted, the bead is pulled along by the kinesin and the nanometer scale displacements, in which the volume with a variety of vesicle pore size is within the range of 10–400 nm.

A light carrier known as an *optical tweezer* can be utilized to exert force on a microscopic object, where a restoring force proportional to the displacement of the microsphere is introduced. An optical trap or "tweezer" is an all-optical noncontact tool, which is grounded on a strongly focused laser beam to trap a dielectric object near the focal point [32]. In principle, the momentum transfer is associated with bending light as the quanta of light energy is proportional to its momentum and in the direction of propagation. The momentum of the light refracts and changes direction when light passes through an object [33]. To conserve the total momentum, an object acquires momentum equal to that lost by the photons, in which the sum of forces can

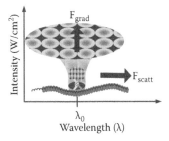

FIGURE 9.3 The optical forces is formed by a potential well, which is generated by a dark soliton.

be separated into two components, the scattering force, in the direction of incident light and the gradient force arising from the intensity gradient pointing toward the center of the beam [34–35]. In this work, a static optical tweezer acting on the object is demonstrated as illustrated in Figure 9.3, where it is used to move the kinesin motor along the microtubule filament. More details of potential wells or dynamic tweezers are found in Deutsch et al. [36]. The trapping phenomenon is categorized into two types as Rayleigh trapping and Mie trapping, which can be used to hold and move microscopic dielectric objects physically with the following details. If the size (d) of the trapped particle is much smaller than the wavelength of the trapping beam (d<<λ), such trapping is known as *trapping in the Rayleigh regime*. The force performing on the particles can be considered as a miniscule dipole immersed in the optical trapping field oscillating at frequency ν. The two forces acting on the particles are known as *dipole scattering* and *Lorentz force* (gradient force). In Mie trapping, the size of the trapped particle is larger than the wavelength (d>>λ). In this chapter, the Rayleigh trapping is described in detail because the molecular motor diameters are much smaller than the carrier signal wavelength. The optical force on the trapped particle can be defined and found in [37]:

$$F = \frac{Q n_m P}{c} \tag{9.1}$$

Q represents the fraction of power utilized to exert force, which is equal to 1, known as dimensionless efficiency, n_m is the refractive index of the suspending medium, *P* is the incident laser power, measured at the tested sample or volume, and *c* is the light speed. For plane wave incident on a perfectly absorbing particle to achieve stable trapping, the radiation pressure must create a stable and three-dimensional equilibrium. As biological specimens are usually immersed in an aqueous medium, the dependence of force *F* on n_m can be used to achieve higher trapping forces. *Q* is therefore the main determinant of trapping force and depends upon the numerical aperture (NA), laser wavelength, laser mode structure, relative index of refraction, light polarization state, and particle geometry. The scattering force is given by [37]:

$$F_{scatt} = n_m \frac{P \langle S \rangle \sigma}{c} \tag{9.2}$$

where

$$\sigma = \frac{8\pi r^2 (kr)^4}{3} \left(\frac{m^2 - 1}{m^2 + 2} \right)^2 \qquad (9.3)$$

Here, P is the incident laser power intensity by optical spin, σ is the scattering cross-section of a Rayleigh sphere with radius r, $\langle S \rangle$ is the time-averaged Poynting vector, n is the index of refraction of the particle, $m = n/n_m$ is the relative index, and $k = 2\pi n_m / \lambda$ is the wave number of the light. The scattering force is proportional to the energy flux and points along the incident light propagation direction.

The time-averaged gradient field is the Lorentz force performing on the dipole induced by the light field given by [37]:

$$F_{grad} = \frac{2\pi\alpha}{cn_m^2} \nabla \langle P \rangle \qquad (9.4)$$

where

$$\alpha = n_m^2 r^3 \left(\frac{m^2 - 1}{m^2 + 2} \right) \qquad (9.5)$$

The gradient force is proportional to the intensity gradient and points up the gradient when $m > 1$. From Equations (9.4) and (9.5), the large gradient force can be obtained by the large depth of the laser beam, in which the stable trapping force requires the gradient force in the $-\hat{z}$ direction against the direction of incident light (dark soliton valley) as shown in Figures 9.5 and 9.6. The gradient field strength can be increased by increasing the laser beam numerical aperture (NA) and decreasing focal spot size. In principle, the potential well is produced among the gaps by two forces to confine the trapped volumes.

9.3   SIMULATION RESULTS

The optical manipulations in many substances are based on an optical dipole interaction model, where a photon from a source can result in a serious effect when light is interacting in a nanoscale structure. Optical dipole is formed by spin manipulation where the dark–bright soliton conversion behaviors can be controlled by using a ring resonator [24]. The PANDA ring resonator is used to form the orthogonal set of dark–bright soliton pair, which can be decomposed into left and right circularly polarized waves as shown in Figure 9.4. The two output light signals relative in phase after coupling into the optical coupler are $\pi/2$. So, the signals coupled into the drop port (Dr) and through port (Th) have a phase difference of π with respect to the input port (In) signal. The output powers at the drop and through ports are given by [24]:

$$|E_d|^2 = \left| \frac{-\kappa_1 \kappa_2 A_{1/2} \Phi_{1/2}}{1 - \tau_1 \tau_2 A\Phi} E_i + \frac{\tau_2 - \tau_1 A\Phi}{1 - \tau_1 \tau_2 A\Phi} E_a \right|^2 \qquad (9.6)$$

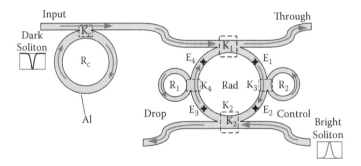

FIGURE 9.4 A schematic diagram of optical tweezer spin generator. (From S. Glomglome et al., Optical Spin Generated by a Soliton Pulse in an Add-Drop Filter for Optoelectronic and Spintronic Use, *Optics & Laser Technolog.*, 44 (5) 1294–1297, 2012.)

$$|E_t|^2 = \left| \frac{\tau_2 - \tau_1 A\Phi}{1 - \tau_1\tau_2 A\Phi} E_i + \frac{-\kappa_1\kappa_2 A_{1/2}\Phi_{1/2}}{1 - \tau_1\tau_2 A\Phi} E_a \right|^2 \qquad (9.7)$$

Here $A_{1/2} = \exp(-\alpha L/4)$ (the half-round-trip amplitude), $A = A_{1/2}^2, \Phi_{1/2} = \exp(j\omega T/2)$ (the half-round-trip phase contribution), and $\Phi = \Phi_{1/2}^2$.

In operation, the orthogonal soliton sets can be generated by using the system shown in Figure 9.4. The optical field is fed into the ring resonator system, where $R_1 = R_2 = 2.5$ µm, $R_{ad} = 30$ µm by using a microring, and $R_c = 20$ µm. To form the initial spin states, the magnetic field is induced by an aluminum plate coupled on *AlGaAs* waveguides for optoelectronic spin-up and spin-down states. The coupling coefficient ratios $\kappa_1{:}\kappa_2$ are 50:50, 90:10, 10:90, and κ_c are 50:50. The system parameters are the ring radii $R_{ad} = 300$ nm, $A_{eff} = 0.25$ µm², $n_{eff} = 3.14$ (for InGaAsP/InP) [24], $\alpha = 0.1$ dB/mm, $\gamma = 0.01$, and $\lambda_0 = 1.45$ µm. The output intensities of spin-injected for transverse electric (TE) and transverse magnetic (TM) fields are generated using a PANDA ring resonator. The optoelectronic fields are generated by a dark-soliton pump based on through port and drop port microring resonators at center wavelength 1.45 µm. The random transverse electric (TE) and transverse magnetic (TM) fields of the solitons corresponding to the left-hand and right-hand photons can be generated and simulated by using the Optiwave and MATLAB programs [24]. The angular momentum of either $+\hbar$ or $-\hbar$ is imparted to the object when a photon is absorbed by an object, where two possible spin states known as optoelectronic spins are exhibited. Many soliton spins can also be generated and detected at through (spin-up) and drop (spin-down) ports of a PANDA ring resonator (Figure 9.5). The optoelectronic spin manipulation generated within a PANDA ring resonator is as shown; the dipole pair with wavelength was varied. Therefore, the output signals of spin optical tweezers by four wavelengths produced in the PANDA ring resonator are obtained at the Th and Dr ports as illustrated in Figure 9.6.

The important aspect of this simulation is that the required dynamic behavior of the tweezers can be obtained by tuning some parameters of the system including add port input signal, coupling coefficient, and ring radius. In this study, this is adapted

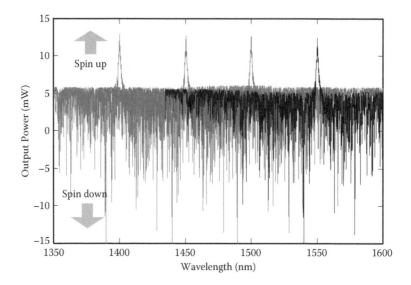

FIGURE 9.5 Photonic spins are generated and confirmed by using a PANDA ring. (From S. Glomglome et al., Optical Spin Generated by a Soliton Pulse in an Add-Drop Filter for Optoelectronic and Spintronic Use, *Optics & Laser Technolog.*, 44 (5) 1294–1297, 2012.)

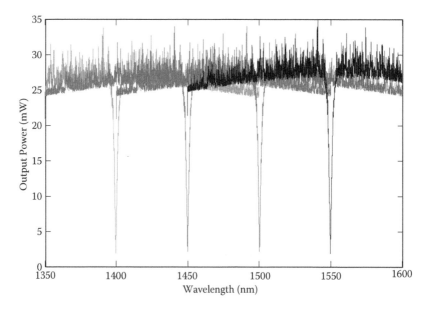

FIGURE 9.6 Optical tweezers with different wavelengths generated by dark–bright soliton pair. (From S. Glomglome et al., Optical Spin Generated by a Soliton Pulse in an Add-Drop Filter for Optoelectronic and Spintronic Use, *Optics & Laser Technolog.*, 44 (5) 1294–1297, 2012.)

by using a microtubule inserting into an InGaAsP/InP waveguide of microtubule filament as shown in Figure 9.7, which consists of two states. First, a device called a *PANDA ring resonator* is used to generate the spin potential well for molecule trapping. Second, the trapped molecules can be moved and rotated to the required destination by the generated potential wells. The kinesin and molecule transportation manipulations are as shown in Figures 9.7 and 9.8, where the waveguide is connected to the microtubule of neuronal cells, in which the microgel/nanogel is used to connect the waveguide all the way through the neuronal cells. The output of optical tweezers from throughput and drop ports with different ring radii are as shown in Figure 9.9, where the different tweezer center wavelengths at 1.4, 1.45, 1.5, 1 nd 1.55 μm can be obtained. Results have shown that the optical tweezers can trap, move, and rotate molecules to the required destination along the waveguide

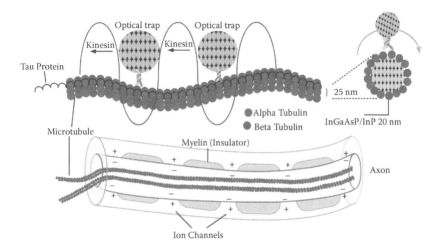

FIGURE 9.7   Molecule movement by molecular motor to the required destinations.

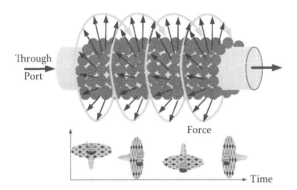

FIGURE 9.8   Molecular motor movement, where the arrows represent the rotation directions along the waveguide surface (through port).

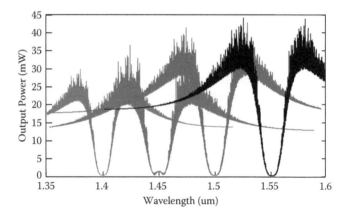

FIGURE 9.9   Optical tweezers (array) for many molecule trappings, where the tweezer center wavelengths are at 1.4, 1.45, 1.5, 1 nd 1.55 μm.

by the plasmonic surfaces. In addition, an array of molecules can be used to trap and rotate many molecules, which can be used to form large-scale molecular spins (molecular motors), in which the different sizes of trapping forces are generated. This allows forming of the different molecular motor sizes, which can be useful for many applications, especially in drug delivery, medical diagnosis, and therapy.

9.4   CONCLUSION

We have demonstrated that the molecular motors using optical tweezers within a modified optical add–drop filter known as a PANDA ring resonator can be generated and performed. By using dark and bright solitons, the orthogonal tweezers can be formed within the system in the same way as photonic spins, in which the rotated tweezers can be generated and controlled. Simulation results have shown that the spin tweezers with the trapped molecules can be rotated (molecular motors), controlled, and moved along the specific waveguide by the plasmonic surfaces, as shown in Figures 9.7 and 9.8, to the required destinations, which is allowed in order to reach the required drug targeting or therapeutic access points. Furthermore, the molecular motor array, that is, many trapped molecules can be generated and moved by using the proposed system, which can be used to form large-scale molecular motors. This is useful for health care technologies, especially drug delivery, medical diagnosis and therapy, in which the required therapeutic targets or accessed points can be controlled and realized.

REFERENCES

1. T. Korten, A. Mansson, and S. Diez, Towards the application of cytoskeletal motor proteins in molecular detection and diagnostic devices, *Curr. Opin. Biotechnol.*, 21 (4) 477–488, 2010.

2. R. D. Vale, The molecular motor toolbox for intracellular transport, *Cell*, 112 (4) 467–480, 2003.
3. Y. Zhang, Cargo transport by several motors, *Phys. Rev. E.*, 83, 011909–18, 2011.
4. Y. Chai, S. Klumpp, M. J. I. Müller, and R. Lipowsky, Traffic by multiple species of molecular motors, *Phys. Rev. E. Stat. Nonlin. Soft Matter Phys.*, 80, 041928–51, 2009.
5. S. Klumpp, Y. Chai, and R. Lipowsky, Effects of the chemomechanical stepping cycle on the traffic of molecular motors, *Phys. Rev. E. Stat. Nonlin. Soft Matter Phys.*, 78, 041909–52, 2008.
6. R. Yokokawa, J. Miwa, M. C. Tarhan, H. Fujita, and M. Kasahara, DNA molecule manipulation by motor proteins for analysis at the single-molecule level, *Anal. Bioanal. Chem.*, 391 (8) 2735–43, 2008.
7. H. J. Silvia and A. G. Ricardo, Exploring mechanochemical processes in the cell with optical tweezers, *Biol. Cell.*, 98 (12) 679–695, 2006.
8. A. Vologodskii, Energy transformation in biological molecular motors, *Physics of Life Rev.*, 3 (2) 119–132, 2006.
9. A. Mogilner, M. Mangel, and R. J. Baskin, Motion of molecular motor ratcheted by internal fluctuations and protein friction, *Phys. Lett. A.*, 237, 297–306, 1998.
10. K. I. Mortensen, L. S. Churchman, J. A. Spudich, and H. Flyvbjerg, Optimized localization analysis for single-molecule tracking and super-resolution microscopy, *Nat. Methods*, 7 (5) 377–381, 2010.
11. L. Jan, O. Bibiana, B. S. Steven, T. J. Ignacio, and B. Carlos, Reversible unfolding of single RNA molecules by mechanical force, *Science*, 292 (5517) 733–737, 2001.
12. H. Nobutaka, N. Yasuko, T. Yosuke, and N. Shinsuke, Kinesin superfamily motor proteins and intracellular transport, *Nat. Rev. Mol. Cell Biol.*, 10 (10) 682–696, 2009.
13. W. Jiaobing and L. F. Ben, Dynamic control of chiral space in a catalytic asymmetric reaction using a molecular motor, *Science*, 331 (6023) 1429–1432, 2011.
14. R. Tucker, P. Katira, and H. Hess, Herding nanotransporters: Localized activation via release and sequestration of control molecules, *Nano. Lett.*, 8 (1) 221–226, 2008.
15. A. Ashkin, J. M. Dziedzic, J. E. Bjorkholm, and S. Chu, Observation of a single-beam gradient force optical trap for dielectric particles, *Opt. Lett.*, 11 (5) 288–290, 1986.
16. L. Huang, S. J. Maerkl, and O. J. F. Martin, Integration of plasmonic trapping in a microfluidic environment, *Opt. Express.*, 17 (8) 6018–6024, 2009.
17. K. C. Neuman and A. Nagy, Single-molecule force spectroscopy: Optical tweezers, magnetic tweezers and atomic force microscopy, *Nat. Methods*, 5, 491–505, 2008.
18. L. J. Mathieu, R. Maurizio, and Q. Romain, Plasmon nano-optical tweezers, *Nat. Photonics*, 5, 349–356, 2011.
19. T. Strick, J. Allemand, V. Croquette, and D. Bensimon, Twisting and stretching single DNA molecules, *Prog. Biophys. Mol. Biol.*, 74 (1–2) 115–140, 2000.
20. M. J. Padgett, J. Courtial, and L. Allen, Light's orbital angular momentum, *Phys. Today*, 57, 35–40, 2004.
21. M. A. Jalil, M. Tasakorn, N. Suwanpayak, J. Ali, and P. P. Yupapin, Nanoscopic volume trapping and transportation using a PANDA ring resonator for drug delivery, *IEEE Trans Nanobioscience.*, 10 (2) 106–112, 2011.
22. M. S. Aziz, N. Suwanpayak, M. A. Jalil, R. Jomtarak, T. Saktioto, J. Ali, and P. P. Yupapin, Gold nanoparticle trapping and delivery for therapeutic applications, *Int. J. Nanomed.*, 7, 11–17, 2012.
23. B. Manfred, Quantum optics: Spin echo with light, *Nat. Photonics*, 4, 347–349, 2010.
24. S. Glomglome, I. Srithanachai, C. Teeka, and P. P. Yupapin, Optical spin generated by a soliton pulse in an add-drop filter for optoelectronic and spintronic use, *Optics & Laser Technolog.*, 44 (5) 1294–1297, 2012.

25. N. Thammawongsa, N. Moongfangklang, S. Mitatha, and P. P. Yupapin, Novel nano-antenna system design using photonic spin in a PANDA ring resonator, *PIER Lett.*, 31, 75–87, 2012.

26. S. Mitatha, N. Moongfangklang, M. A. Jalil, N. Suwanpayak, J. Ali, and P. P. Yupapin, Multi-access drug delivery network and stability, *Int. J. Nanomed.*, 6, 1757–1764, 2011.

27. R. Lipowsky, J. Beeg, R. Dimova, S. Liepelt, S. Klumpp, M. J. I. Müller, and A. Valleriani, Active bio-systems: From single motor molecules to cooperative cargo transport, *Biophys. Rev. Lett.*, 4 (1–2) 77–137, 2009.

28. L. C. Kapitein, M. A. Schlager, M. Kuijpers, P. S. Wulf, M. van Spronsen, F. C. MacKintosh, and C. C. Hoogenraad, Mixed microtubules steer dynein-driven cargo transport into dendrites, *Curr. Biol.*, 20 (4) 290–299, 2010.

29. C. Veigel and C. F. Schmidt, Moving into the cell: Single-molecule studies of molecular motors in complex environments, *Nat. Rev. Mol. Cell. Bio.*, 12, 163–176, 2011.

30. J. Kierfeld, K. Frentzel, P. Kraikivski, and R. Lipowsky, Active dynamics of filaments in motility assays, *Eur. Phys. J.*, 157 (1) 123–133, 2008.

31. K. Aathavan, A. T. Politzer, A. Kaplan, J. R. Moffitt, Y. R. Chemla, S. Grimes, P. J. Jardine, D. L. Anderson, and C. Bustamante, Substrate interactions and promiscuity in a viral DNA packaging motor, *Nature*, 461, 669–674, 2009.

32. G. Volpe, M. J. Padgett, J. Molloy, and D. McGloin, Optical tweezers: Methods and applications, *Contemporary Phys.*, 52, 2011.

33. M. P. Sheetz, L. Wilson, and P. Matsudaira, *Laser Tweezers in Cell Biology*, San Diego, CA: Academic Press, 55, 1997.

34. A. Ashkin, Forces of a single-beam gradient force on a dielectric sphere in the ray of optic regime, *Biophys. J.*, 61 (2) 569–582, 1992.

35. A. Ashkin, K. Schutze, J. M. Dziedzic, U. Euteneuer, and M. Schliwa, Force generation of organelle transport measured *in vivo* by an infrared laser trap, *Nature*, 348, 346–348, 1990.

36. I. H. Deutsch, P. M. Alsing, J. Grondalski, S. Ghose, P. S. Jessen, and D. L. Haycock, Quantum transport in magneto-optical double-potential wells, *J. Opt. B: Quant. and Semiclass. Opt.*, 2, 633–644, 2000.

37. K. Svoboda and S. M. Block, Biological applications of optical forces, *Annu. Rev. Biophys. Biomol. Struct.*, 23, 247–283, 1994.

10 Cancer Cell Treatment by Short Pulse Laser

10.1 INTRODUCTION

Ultrashort laser pulse thermal-based killing of abnormal cells such as cancer cells targeted using absorbing nanoparticles is becoming an interesting subject in nanophotonics and medical treatments [1–3]. Thermal treatment using photo-absorption causes generation of heat from optical energy, which has many beneficial advantages compared to other treatment methodologies and cancer therapies such as surgery and radiotherapy. A wide range of nanomaterials with strong optical absorption in the near infrared (NIR) such as gold nanoparticles [4], nanoshells [5], or carbon nanotubes [6–8] can be used as photothermal. Laser energies of ultrashort pulses can be concentrated on a small biological mass filled with metallic nanoparticles. Therefore, an ultrashort laser pulse shoots the photons toward the target, where it can be absorbed by free electrons within the metal and transferred to the lattice subsystem, and finally to the surrounding medium. The knowledge of laser interactions with nanoparticles has required specific models for each case, dependent upon the laser pulse duration. The dual-temperature model for the ultrashort laser pulse mode is readily available [9–10].

Ultrashort pulses, specifically femtosecond and picosecond, are appropriate for medical treatment applications. Free electrons of nanoparticle materials such as gold are able to absorb energy, quickly get to high temperatures, and transfer thermal energy to the lattice. The delay time of electron cooling and lattice heating is on the order of femtoseconds and picoseconds, respectively [11–12]. Two appropriate models of a one-temperature model (OTM) and a two-temperature model (TTM) are well described, where OTM is operating uniformly and there is a suitable heating approximation for nanoparticle heating in the femtosecond and picosecond regimes. OTM provides an effective modeling method for further nanomedicine research to explore [13]. Nanoparticles in blood and tumor are comparable to the case of water surrounding a medium because the thermodynamic properties of those media are very close to each other. Fat has low-thermal characteristics compared to water; therefore, higher overheating of the gold nanoparticle can be seen at the same energy level.

The thermal equilibrium among the excited electrons is within 100–175 femtoseconds. Electron cooling time due to the coupling to lattice is about 10 picoseconds. In the case of the OTM model, the heat flows from the nanoparticle into the surrounding medium and becomes noticeable after 200 picoseconds. Transferred energy from laser pulses to the medium uses a wavelength, which can be strongly absorbed by the particles but not by cells. Gold nanoparticles can strongly absorb visible light, where it has a distinct absorption peak near 520 nm [14].

This technique causes localized tumor damage without harmful effects on surrounding healthy tissue. Absorption of laser heat occurs at longer wavelengths for most biotissues with relatively fast treatment. Promising results are obtained with the use of gold nanoshells, which are heated in the range of around 800 nm wavelength using continuous laser radiation [15]. To generate a spectrum of light over a broad range, an optical soliton pulse is recommended as a powerful laser pulse that can be used to generate chaotic filter characteristics when propagating within nonlinear microring resonators (MRRs) [16].

Using this technique, the capacity of the transmission data can be secured and increased when chaotic packet switching is employed [17]. In this study, we propose a system that can be used for extremely short pulses in the range of pico- and femtosecond of soliton used for abnormal cell ablation such as cancer or tumor.

Ring resonators can be used to generate ultrashort pulses in the nonlinear regime, where use of a soliton laser becomes an interesting subject [18]. High optical output signals of the rings system are of benefit. There are two techniques that can be used to generate soliton laser pulses. First, Pornsuwancharoen et al. showed that soliton pulses can be obtained using a pumped soliton pulse inside a ring system where large amplified signals are being achieved [19]. Second, a Gaussian soliton can be generated in a simple system arrangement. This application becomes an attractive tool in the area of nanomedicine applications. However, there is much research which is reported in both theoretical and experimental work using a bright soliton pulse for soliton study [20].

Such an extremely narrow signal in the range of pico- and femtoseconds soliton pulses can be used for many applications in medical ablation. In this work a monochromatic laser source such as a bright soliton pulse with central wavelength of 0.6 nm and 1550 μm is used to generate high-capacity ultrashort soliton signals within a nonlinear fiber ring resonator.

10.2   THEORETICAL MODELING

The schematic diagram of the proposed system is shown in Figure 10.1. A soliton pulse with 20 ns pulse width, peak power at 500 mW is input into the system. Suitable ring parameters are used, for instance, ring radii $R_1 = 10$ μm, $R_2 = 5$ μm, and $R_3 = 4$ μm. In order to make the system associate with the practical device, the selected parameters of the system are fixed to $\lambda_0 = 0.6$ μm and 1.55 μm, $n_0 = 3.34$ (InGaAsP/InP), $A_{eff} = 0.50, 0.25$ μm^2, and 0.12 μm^2 for different radii of microring resonators, respectively, $\alpha = 0.5$ dBmm^{-1}, $\gamma = 0.1$. The coupling coefficient (kappa, κ) of the microring resonator ranged from 0.50 to 0.975 [21].

The bright soliton pulse is introduced into the proposed system, where the input optical field (E_{in}) of the bright soliton pulse is given by [22],

$$E_{in} = A \sec h\left[\frac{T}{T_0}\right]\exp\left[\left(\frac{z}{2L_D}\right) - i\omega_0 t\right] \qquad (10.1)$$

A and z are the optical field amplitude and propagation distance, respectively. T is a soliton pulse propagation time in a frame moving at the group velocity, $T = t - \beta_1 \times z$,

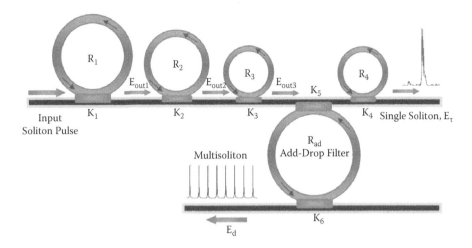

FIGURE 10.1 Schematic diagram of single and multiple ultrashort pulse generation for thermal-based killing of abnormal cells.

where β_1 and β_2 are the coefficients of the linear and second-order terms of the Taylor expansion of the propagation constant. $L_D = T_0^2/|\beta_2|$ is the dispersion length of the soliton pulse. The frequency shift of the soliton is ω_0. This solution describes a pulse that keeps its temporal width invariance as it propagates, and thus is called a *temporal* and *spatial soliton*. When soliton peak intensity $\left(|\beta_2/\Gamma T_0^2|\right)$ is given, then T_0 is known. For the soliton pulse in the microring device, a balance should be achieved between the dispersion length (L_D) and the nonlinear length ($L_{NL} = 1/\Gamma\varphi_{NL}$), where $\Gamma = n_2 \times k_0$, is the length scale over which dispersive or nonlinear effects make the beam become wider or narrower, hence $L_D = L_{NL}$. When light propagates within the nonlinear medium, the refractive index (n) of light within the medium is given by

$$n = n_0 + n_2 I = n_0 + \left(\frac{n_2}{A_{eff}}\right) P \tag{10.2}$$

where n_0 and n_2 are the linear and nonlinear refractive indexes, respectively. I and P are the optical intensity and optical power, respectively. The effective mode core area of the device is given by A_{eff}. For the MRR, the effective mode core areas range from 0.50 to 0.10 μm^2 [23]. When a soliton pulse is input and propagated within a microring resonator as shown in Figure 10.1, the resonant output is formed, thus, the normalized output of the light field is the ratio between the output and input fields $E_{out}(t)$ and $E_{in}(t)$ in each round trip, which can be expressed as;

$$\left|\frac{E_{out}(t)}{E_{in}(t)}\right|^2 = (1-\gamma)\left[1 - \frac{(1-(1-\gamma)x^2)\kappa}{\left(1-x\sqrt{1-\gamma}\sqrt{1-\kappa}\right)^2 + 4x\sqrt{1-\gamma}\sqrt{1-\kappa}\sin^2\left(\frac{\Phi}{2}\right)}\right] \tag{10.3}$$

Here, κ is the coupling coefficient, and $x = \exp(-\alpha L/2)$ represents a round-trip loss coefficient, $\Phi_0 = kLn_0$ and $\Phi_{NL} = kLn_2|E_{in}|^2$ are the linear and nonlinear phase shifts, $k = 2\pi/\lambda$ is the wave propagation number in a vacuum, where L and α are a waveguide length and linear absorption coefficient, respectively. The optical outputs of a ring resonator add/drop filter system can be expressed by Equations (10.4) and (10.5) [24].

$$|E_t|^2 = |E_{outR3}|^2 \times \frac{(1-\kappa_1) - 2\sqrt{1-\kappa_1} \cdot \sqrt{1-\kappa_2} e^{\frac{\alpha}{2}L} \cos(k_n L) + (1-\kappa_2)e^{-\alpha L}}{1 + (1-\kappa_1)(1-\kappa_2)e^{-\alpha L} - 2\sqrt{1-\kappa_1} \cdot \sqrt{1-\kappa_2} e^{\frac{\alpha}{2}L} \cos(k_n L)} \qquad (10.4)$$

$$|E_d|^2 = |E_{outR3}|^2 \times \frac{\kappa_1 \kappa_2 e^{\frac{\alpha}{2}L}}{1 + (1-\kappa_1)(1-\kappa_2)e^{-\alpha L} - 2\sqrt{1-\kappa_1} \cdot \sqrt{1-\kappa_2} e^{\frac{\alpha}{2}L} \cos(k_n L)} \qquad (10.5)$$

where E_t and E_d are the optical outputs of the through and drop ports, respectively, while the output power from the third ring of the MRR's system is shown by E_{outR3}. The propagation constant is given by $\beta = kn_{eff}$, where the effective refractive index of the waveguide is represented by n_{eff}. The circumference of the add/drop ring is $L = 2\pi R$ where R is the radius. When light travels in an MRR system at the point where the ring becomes close to the straight waveguide, only the resonance wavelength of light is coupled into the ring. By using the particular parameters of the add/drop device, the chaotic noise cancellation can be implemented. The coupling coefficients of the add/drop filter have been shown by κ_4 and κ_4. The microring resonator loss and the fractional coupler intensity loss are $\alpha = 0.5$ dBmm^{-1} and $\gamma = 0.1$, respectively, while the nonlinear refractive index can be neglected for an add/drop device [25].

10.3   RESULTS AND DISCUSSION

In operation, the large bandwidth within the microring device can be generated by using a soliton pulse input into the nonlinear MRR shown in Figure 10.1, where the required signals can perform the generation of ultrashort soliton pulses. The nonlinear refractive index is $n_2 = 2.2 \times 10^{-17}$ m^2/W. As shown in Figure 10.2, the input signal is chopped (sliced) into smaller signals spreading over the spectrum, which shows that the large bandwidth is formed within the first ring device. The compressed bandwidth with smaller group velocity is obtained within the ring R_2. The amplified gain is obtained within the R_3 microring device. The temporal soliton pulse can be formed by using the constant gain condition, where a small group velocity is seen. Attenuation of the optical power within a microring device is required in order to keep the constant output gain. In this case, the time belongs to each ring resonator. Similarly, the spatial soliton is obtained as shown in Figure 10.3.

Results of femtosecond pulse generation can be seen from Figure 10.4, where the radius of the rings has been selected to $R_1 = 17$ μm, $R_2 = 13$ μm, $R_3 = 7$ μm, $R_4 = 2$ μm. The center wavelength of the input bright soliton is $\lambda = 0.6$ μm. Spatial soliton

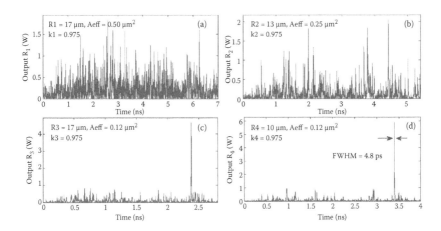

FIGURE 10.2 Results obtained when a temporal soliton is localized within a microring device with 20,000 round trips, where (a) Chaotic signals from R_1, (b) Chaotic signals from R_2, (c) Trapping of temporal soliton, and (d) Trapped temporal soliton with FWHM of 4.8 ps.

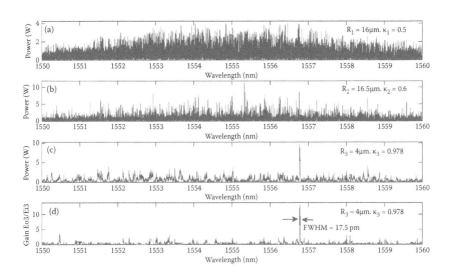

FIGURE 10.3 Results of trapping spatial soliton within a microring device with 20,000 round trips, where (a) Chaotic signals from R_1, (b) Chaotic signals from R_2, (c) Trapping of spatial soliton, and (d) Trapped spatial soliton with FWHM of 17.5 pm.

pulse with full width at half maximum (FWHM) of 0.34 nm can be generated at λ = 555 nm shown in Figure 10.5d, where the steps of the filtering process are shown in Figures 10.5a,b,c. Figure 10.6 shows the results when temporal and spatial optical soliton pulses are localized within a microring device and add/drop filter system with 20,000 round trips, where optical ultrashort temporal soliton of FWHM = 83 fs is generated. Here, the ring radii are R_1 = 10 μm, R_2 = 5 μm, R_3 = 4 μm, R_4 = 4

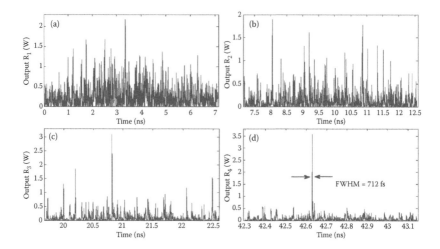

FIGURE 10.4 Results obtained when a temporal soliton is localized within a microring device with 20,000 round trips, where (a) Chaotic signals from R_1, (b) Chaotic signals from R_2, (c) Trapping of temporal soliton, and (d) Trapped temporal soliton with FWHM of 712 fs.

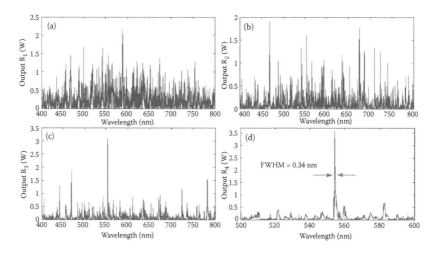

FIGURE 10.5 Results of trapping spatial soliton within a microring device with 20,000 round trips, where (a) Chaotic signals from R_1, (b) Chaotic signals from R_2, (c) Trapping of spatial soliton, and (d) Trapped spatial soliton with FWHM of 0.34 nm.

μm, and $R_{ad} = 200$ μm with coupling coefficient of $\kappa_1 = 0.3$, $\kappa_2 = 0.5$, $\kappa_3 = 0.7$, $\kappa_4 = 0.9$, $\kappa_5 = 0.1$, and $\kappa_6 = 0.1$. Figure 10.7 shows the results when a localized ultrashort temporal optical soliton with 84 fs FWHM is generated, where multitemporal optical solitons with FWHM = 140 ps and FSR = 3.6 ns are generated. The multispatial optical solitons have FWHM = 40 pm and FSR = 1.45 nm, where the ring radius

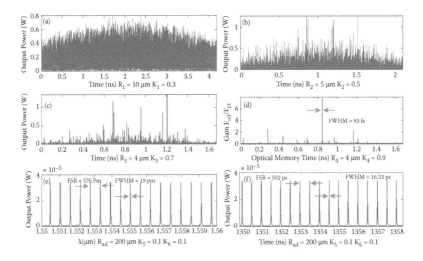

FIGURE 10.6 Results of temporal and spatial soliton generation, where (a) Chaotic signals from R_1, (b) Chaotic signals from R_2, (c) Filtering signals, (d) Trapped temporal soliton with FWHM of 83 fs, (e) Spatial soliton with FSR = 576 pm and FWHM = 19 pm, and (f) Temporal soliton with FSR = 502 ps and FWHM = 16.53 ps.

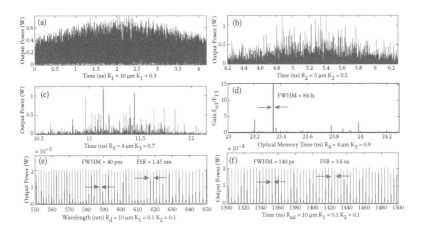

FIGURE 10.7 Results of temporal and spatial soliton generation, where (a) Chaotic signals from R_1, (b) Chaotic signals from R_2, (c) Filtering signals, (d) Trapped temporal soliton with FWHM of 84 fs, (e) Spatial soliton with FSR = 1.45 nm and FWHM = 40 pm, and (f) Temporal soliton with FSR = 3.6 ns and FWHM = 140 ps.

of the add/drop filter system is $R_{ad} = 10$ μm and the central wavelength of the input bright soliton power has been selected to $\lambda = 0.6$ μm. Precise ablation that is possible with pico- and femtosecond lasers makes them ideal candidates for high-precision surgery [26]. In the case of minimizing thermal effects in the area surrounding the abnormal target volume, femtosecond laser pulses can be used to selectively

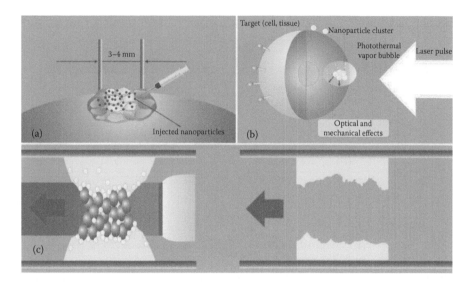

FIGURE 10.8 Schematic of cancer cell disruption, where (a) Gold nanoparticles are injected to the target site, (b) Gold nanoparticles interact with ultrashort laser pulse, where photothermal bubbles are generated around gold nanoparticles, and (c) Photothermal bubbles mechanically disrupt and remove the tissue, thus creating a channel.

eliminate cellular components in individual cells, making them a powerful research tool. Interaction of ultrashort laser pulse with gold nanoparticles, which leads to transferring generated heat to abnormal cells such as tumor, cancer, or plaque is shown in Figure 10.8. Effects of the laser–nanoparticle interaction with the cancer cells is shown in Figure 10.8c. The optical and mechanical effects of the bubble can be controlled through the laser parameters in order to tune the bubble to a diagnostic or therapeutic task [27]. Gold nanoparticles are the most promising candidates for photothermolysis since they are strong absorbers, photostable, and have adjustable optical properties. Therefore, gold nanoparticles in the size range of 2–250 nm form a cluster around a target molecule where they are excited with a short laser pulse acting as a heat source, thus generating an intracellular photothermal vapor bubble.

10.4 CONCLUSION

We have proposed an interesting concept in which ultrashort temporal and spatial soliton pulses in the pico- and femtosecond range can be generated using the MRR system. We have shown that a large bandwidth of the arbitrary soliton pulses can be generated and compressed within a micro-waveguide. The chaotic signal generation using a soliton pulse in the nonlinear microring resonators has been presented. Selected ultrashort light pulses can be localized and used to perform thermal-based killing of abnormal cells, tumor, and cancer. Laser pulses interact with gold nanoparticles and generate optical and mechanical heating effects, which cause an increase of temperature around the abnormal cell leading to clean removal of the target.

REFERENCES

1. C. M. Pitsillides, E. K. Joe, X. Wei, R. Anderson, and C. P. Lin, Nanometer scale quantum thermometry in a living cell, *Biophysical Journal,* 84, 4023, 2003.
2. V. Zharov, R. Letfullin, and E. Galitovskaya, Microbubbles-overlapping mode for laser killing of cancer cells with absorbing nanoparticle clusters, *Journal of Physics D: Applied Physics,* 38, 2571, 2005.
3. R. R. Letfullin, C. Joenathan, T. F. George, and V. P. Zharov, Laser-induced explosion of gold nanoparticles: Potential role for photothermalysis of cancer cells, *Nanomedicine,* 1, 473, 2006.
4. G. Von Maltzahn, J. H. Park, A. Agrawal, N. K. Bandaru, S. K. Das, M. J. Sailor, and S. N. Bhatia, Guided delivery of polymer therapeutics using plasmonic photothermal therapy, *Cancer Research,* 69, 3892, 2009.
5. L. R. Hirsch, R. Stafford, J. Bankson, S. Sershen, B. Rivera, R. Price, J. Hazle, N. Halas, and J. West, Narrow band imaging of tumors using gold nanoshells, *Proceedings of the National Academy of Sciences,* 100, 13549, 2003.
6. X. Liu, H. Tao, K. Yang, S. Zhang, S. T. Lee, and Z. Liu, Optimization of surface chemistry on single-walled carbon nanotubes for *in vivo* photothermal ablation of tumors, *Biomaterials,* 32, 144, 2011.
7. S. Ghosh, S. Dutta, E. Gomes, D. Carroll, R. D'Agostino Jr, J. Olson, M. Guthold, and W. H. Gmeiner, Increased heating efficiency and selective thermal ablation of malignant tissue with DNA-encased multiwalled carbon nanotubes, *ACS Nano,* 3, 2667, 2009.
8. H. K. Moon, S. H. Lee, and H. C. Choi, *In vivo* and *in vitro* with EGF-directed carbon nanotube-based drug delivery, *ACS Nano,* 3, 3707, 2009.
9. J. Wu, C. Wang, and G. Zhang, Ultrafast optical response of the Au–BaO thin film stimulated by femtosecond pulse laser, *Journal of Applied Physics,* 83, 7855, 1998.
10. B. Chichkov, C. Momma, S. Nolte, F. Von Alvensleben, and A. Tünnermann, Development of niobium-based superconducting junctions, *Applied Physics A: Materials Science & Processing,* 63, 109, 1996.
11. N. Inogamov, V. Zhakhovskii, S. Ashitkov, V. Khokhlov, Y. V. Petrov, P. Komarov, M. Agranat, S. Anisimov, and K. Nishihara, Nanomodification of gold surface by picosecond soft x-ray laser pulse, *Applied Surface Science,* 255, 9712, 2009.
12. B. Krenzer, A. Hanisch, A. Duvenbeck, B. Rethfeld, and M. H. Hoegen, Heat transport in nanoscale heterosystems: A numerical and analytical study, *Journal of Nanomaterials,* 13, 2008.
13. R. R. Anderson and J. A. Parrish, Selective photothermolysis: Precise microsurgery by selective absorption of pulsed radiation, *Science,* 220, 524, 1983.
14. G. Mie, Beiträge zur optik trüber medien, speziell kolloidaler metallösun, *Annalen der Physik,* 330, 377, 1908.
15. D. P. O'Neal, L. R. Hirsch, N. J. Halas, J. D. Payne, and J. L. West, Photo-thermal tumor ablation in mice using near infrared-absorbing nanoparticles, *Cancer Letters,* 209, 171, 2004.
16. P. Yupapin, N. Pornsuwancharoen, and S. Chaiyasoonthorn, Attosecond pulse generation using the multistage nonlinear microring resonators, *Microwave and Optical Technology Letters,* 50, 3108, 2008.
17. I. Amiri, K. Raman, A. Afroozeh, M. Jalil, I. Nawi, J. Ali, and P. Yupapin, Analytical treatment of parametric effects in a ring resonator, *Procedia Engineering,* 8, 360, 2011.
18. E. A. Gibson, X. Zhang, T. Popmintchev, A. Paul, N. Wagner, A. Lytle, I. P. Christov, M. M. Murnane, and H. C. Kapteyn, Extreme nonlinear optics: Attosecond photonics at short wavelength, *IEEE Journal of Selected Topics in Quantum Electronics,* 10, 1339, 2004.

19. N. Pornsuwancharoen, J. Kumbun, and P. Yupapin, Packet switching start–stop bits generation based on bifurcation behavior of light in a micro ring resonator, *Advanced Materials Research*, 55, 505, 2008.

20. H. G. Purwins, H. Bödeker, and S. Amiranashvili, Dissipative solitons, *Advances in Physics*, 59, 485, 2010.

21. J. Toulouse, Optical nonlinearities in fibre: Review, recent examples and systems applications, *Journal of Lightwave Technology*, 23, 3625, 2005.

22. P. Giri, K. Choudhary, A. S. Gupta, A. Bandyopadhyay, and A. R. McGurn, Solitons in a composite multiferroic chain, *Physical Review B*, 84, 155429, 2011.

23. K. Hiruma, K. Tomioka, P. Mohan, L. Yang, J. Noborisaka, B. Hua, A. Hayashida, S. Fujisawa, S. Hara, and J. Motohisa, Catalytic growth of zinc oxide nanowires by vapor transport, *Journal of Nanotechnology*, 29, 2012.

24. A. Afroozeh, I. S. Amiri, M. A. Jalil, M. Kouhnavard, J. Ali, and P. P. Yupapin, Multi soliton generation for enhanced optical communication, *Applied Mechanics and Materials*, 83, 136, 2011.

25. A. F. G. F. Filho, J. R. R. Sousa, F. Lima, W. Fraga, G. Guimarães, J. W. M. Mendonça, and A. S. B. Sombra, A performance study of a nonlinear all fibre Michelson interferometer, add-drop multiplexer, based in fibre Bragg grating mirrors, *Optical and Quantum Electronics*, 40, 525, 2008.

26. V. Wieger, J. Wernisch, and E. Wintner, Novel oral applications of ultra-short laser pulses, *SPIE*, 6460, 64600B, 2007.

27. R. Zhao, Z. C. Liang, R. Q. Xu, J. Lu, and X. W. Ni, Adaptive total variation denoising based on difference curvature, *Nanjing Youdian Daxue Xuebao (Ziran Kexue Ban)*, 29, 83, 2009.

11 Microsurgery

11.1 INTRODUCTION

Nanotechnology devices are increasingly being used for medical applications. The use of nanorobots is an advance in biomedical involvement with minimal invasive surgery [1], continuous patient data acquisition [2], neurosurgery preparation [2], cancer stage diagnosis [3], medical monitoring [4], blood pressure control [5], and improved treatment efficiency [6]. The parameters for medical nanorobot architecture and control require a technology background that leads to the manufacture of the hardware for molecular machines [7]. The nanorobot architecture for medical use must include the necessary devices to operate in the human body workspace with different temperatures, and electromagnetic and chemical gradients in the cell site [8]. To achieve this aim, energy supply, data processing, and data transmission capabilities can be used to advance embedded integrated circuits derived from nanotechnology and very large-scale integrated (VLSI) circuit design [9]. Developments in biomolecular research [10,11] have demonstrated positively the feasibility of nanorobots.

Basically, nanorobots are a controllable machine on a molecular scale, which is composed of nanoscale components. Actually, the main goal in the field of molecular machines is to use various biological elements functioning at the cellular level to create motion, force, or a signal as machine components. All the different components are assembled together in suitable proportion and orientation to form nanodevices, and are able to manipulate and apply force [12]. The advantage of using nature's machine components such as proteins and DNA are high efficiency and reliability, which can operate as motors, mechanical joints, transmission elements, or sensors. Thus, the latest bio-nanorobot design was proposed based jointly on biotechnology and nanoelectronics [13]. Bio-nanorobots have the advantage of delivering an on-site surgeon inside the human body through blood vessels without any side effect.

The manipulation concept to transport a nanoparticle was first proposed by Ashkin in 1987, in which the optical trap mechanism called *optical tweezers* for manipulating biological objects was demonstrated [14]. This started the new era of molecular motor control by light. In biological applications for optical trapping and manipulation, it is possible to remotely apply controlled forces on living cells, internal parts of cells, and to large biological molecules without inflicting detectable optical damage. There are several studies of different dynamic and biomolecular processes, ranging from individual macromolecules such as proteins, DNA, and RNA unfolding under force on molecular motor translocation and exerting forces [15]. Of particular interest is the single-molecule optical trap experiment that provides novel insights on the mechanism of nucleic acid translocation and related processes for trapping nanoparticles [16]. Another large area is the technique of optically assisted *in vitro* fertilization, which is being studied as well as problems in cell recognition, cell fusion, chromosome motion during cell division, and the effects of gravity on plant roots. Therefore, new research techniques are needed to modify molecular motors to bio-nanorobots, which have many advantages for medical applications.

A new research vision based on optical spin is becoming increasingly important [17,18]. There are many optical spin applications such as nanocommunication networks, security and sensing networks, long-distance optical transport, quantum computers, and communication. In addition, spin mechanisms using bright and dark soliton conversion behaviors was demonstrated by Yupapin et al. [19]. It consists of a modified add-drop optical filter known as a PANDA ring resonator, which is capable of generating dynamic optical tweezers to trap the nanoparticles. An improved new optical trapping design to transport gold nanoparticles using a PANDA ring resonator system was reported by Aziz et al. [20]. The intense optical fields in the form of a dark soliton controlled by soliton pulses are used to trap and transport nanoscopic volumes of matter to the desired destination via an optical waveguide. In principle, optical tweezers are generated from the forces exerted by intensity gradients in the strongly focused beams of light to trap and move the microscopic volumes of matter. This is done by a combination of forces induced by the interaction between photons, due to the photon scattering effects. In application, the field intensity can be adjusted and tuned to generate the desired gradient field. The scattering force can then form the suitable trapping force.

There is much literature on optical tweezers and nanorobot manipulation, but there is less research investigation about optical control for nanorobots. In this chapter, the dynamic behavior of the tweezers is demonstrated in the same way as the optical spin to rotate the trap volume [21] is used as a spin power for a nanorobot motor. It can transport nanorobots for targeted surgical and treatment applications. Finally, a new concept of nanorobot control using dynamic optical tweezers within a modified optical add–drop filter called a *PANDA ring resonator* is established. In principle, the optical tweezers are generated by a dark and bright soliton pair corresponding to the left-hand and right-hand rotating solitons. The carrier signals in the form of optical vortices or potential wells can be used to rotate nanorobots and for transport using the dynamic optical tweezers. The merit of such a system is the highly stable signals with no fluctuations over a certain period of time along a fiber optic liquid core, which is useful for many applications, especially in medical diagnosis and therapy.

11.2 NANOROBOT TRANSPORTATION

The optical tweezers are formed by spin manipulation [21]. The PANDA ring resonator is used to generate the orthogonal set of dark–bright soliton pairs, which can be decomposed into left and right circularly polarized waves. The two output signals relative phase after coupling into the optical coupler is $\pi/2$. Hence, the signals coupled into the through port (*Th*) and drop port (*Dr*) have a π phase difference. The input and control fields at the input and control ports (*Ct*) are formed by the dark and bright optical solitons as described in Sarapat et al. [21].

In principle, the molecular transportation system in which orthogonal soliton sets can be generated by modified add-drop are shown in Figure 11.1. The optical field is fed into the ring resonator system, where $R_1 = R_2 = 5$ μm, $R_{ad} = 30$ μm by using

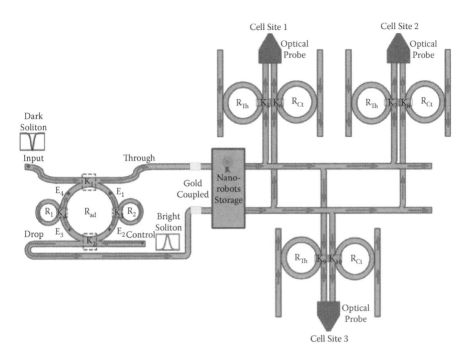

FIGURE 11.1   A schematic diagram of a molecular transportation network.

a microring, and R_c = 18 μm. To form the initial spin states, the magnetic field is induced by gold coupled on AlGaAs waveguides for optoelectronic spin-up and spin-down states.

The coupling coefficient ratios κ_1:κ_2 are 50:50, 90:10, 10:90, and κ_c are 50:50. The system parameters are A_{eff} = 0.25 μm², n_{eff} = 3.14 (for InGaAsP/InP), α = 0.1 dB/mm, γ = 0.01, and λ_0 = 1.55 μm. The optoelectronic fields are generated by a dark-soliton pump based on through port and drop port microring resonator at center wavelength 1.45 μm. The output intensities of spin-injected for transverse electric (TE) and transverse magnetic (TM) fields are generated by using a PANDA ring resonator. The solitons' TE and TM fields corresponding to the left-hand and right-hand photons can be generated and simulated by using the Optiwave and MATLAB program. The angular momentum of either +ħ or –ħ is imparted to the object when a photon is absorbed by an object. Figure 11.2 illustrates a gold coupling model in which the signal is passed through and traps nanorobots in the storage device. The optoelectronic spin manipulation generates the potential well to trap nanorobots as shown in Figure 11.3. Therefore, the output signals of the spin optical tweezers are obtained at the through and drop ports as illustrated in Figure 11.3.

Optical trapping has important implications for a deeper understanding of nanorobot transport in our system. Optical traps or tweezers are an all-optical noncontact tool, which is grounded on a strongly focused laser beam to trap a dielectric object near the focal point [22]. The trap applies a force proportional to the displacement of the microsphere. In theory, the momentum transfer is associated with bending light as

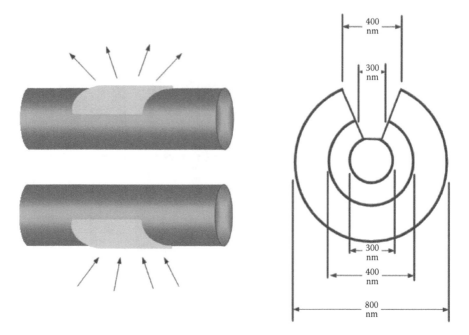

FIGURE 11.2   Fiber optic with gold coupling design for optical spin tweezers.

the quanta of light carries energy proportional to its momentum and in the direction of propagation. The light momentum changes the direction when light passes through an object. To conserve the total momentum, an object obtains momentum equal to a photon lost. The static optical tweezers act on the object, as demonstrated in Figure 11.4, which is used to move bio-nanorobots along a microtubule filament. Potential well or dynamic tweezers were reported in Deutsch et al. [23] in which the sum of forces can be separated into two components, the scattering force, in the direction of incident light, and the gradient force arising from the intensity gradient pointing toward the center of the beam. These two forces can be used to hold and move microscopic dielectric objects physically as shown in Figure 11.5. This trapping phenomenon is categorized into two types as Rayleigh trapping and Mie trapping. If the size of the trapped particle is much smaller than the wavelength of the trapping beam (d<<λ), such trapping is known as *trapping in the Rayleigh regime*. The force acting on the particles can be considered as a miniscule dipole immersed in the optical trapping field oscillating at frequency ν. The two forces acting on the particles are known as *dipole scattering* and *Lorentz force* (gradient force). In Mie trapping, the size of the trapped particle is larger than the wavelength (d>>λ).

In this chapter, the Rayleigh trapping is described in detail because the molecular motor diameters are much smaller than the carrier signal's wavelength. The optical forces, which perform on the trapped particles, can be defined as [24],

$$F = \frac{Q n_m P}{c} \qquad (11.1)$$

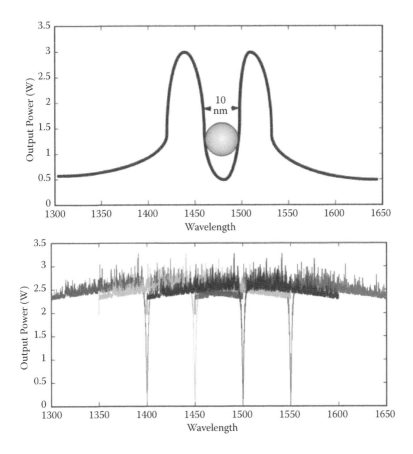

FIGURE 11.3 Optical tweezers generated by dark solitons.

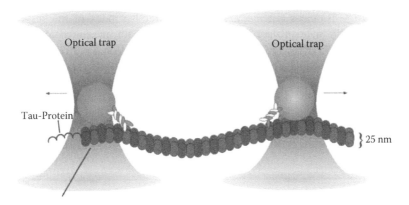

FIGURE 11.4 Optical tweezers for nanorobot movement by optical trapping along the microtubule.

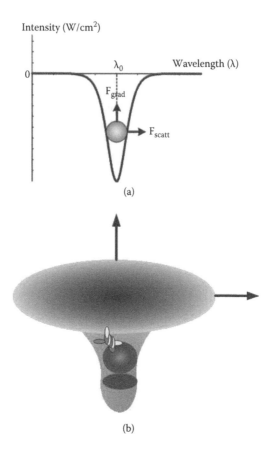

FIGURE 11.5   (a) The optical forces and (b) particle (nanorobot) is trapped by a potential well, which is formed by a dark soliton.

Here, Q represents the fraction of power utilized to exert force, which is equal to 1, known as dimensionless efficiency, n_m is the refractive index of the suspending medium, P is the incident laser power from optical spin, measured at the tested sample, and c is the light speed. For plane wave incident on a perfectly absorbing particle to achieve stable trapping, the radiation pressure must create a stable and three-dimensional equilibrium. As biological specimens are usually immersed in an aqueous medium, the dependence of force F on n_m can be used to achieve higher trapping forces. Q is the main determinant of trapping force and depends upon the numerical aperture (NA), laser wavelength, laser mode structure, relative index of refraction, light polarization state, and particle geometry. The scattering force is given by [24],

$$F_{scatt} = n_m \frac{P \langle S \rangle \sigma}{c} \tag{11.2}$$

where

$$\sigma = \frac{8\pi r^2 (kr)^4}{3} \left(\frac{m^2 - 1}{m^2 + 2} \right)^2 \qquad (11.3)$$

Here, P is the incident laser power intensity by optical spin, σ is the scattering cross-section of a Rayleigh sphere with radius r, $\langle S \rangle$ is the time-averaged Poynting vector, n is the index of refraction of the particle, $m = n/n_m$ is the relative index, and $k = 2\pi n_m/\lambda$ is the wave number of the light. The scattering force is proportional to the energy flux and points along the incident light propagation direction.

The time-averaged gradient field is the Lorentz force performing on the dipole induced by the light field given by [24],

$$F_{grad} = \frac{2\pi\alpha}{cn_m^2} \nabla \langle P \rangle \qquad (11.4)$$

where

$$\alpha = n_m^2 r^3 \left(\frac{m^2 - 1}{m^2 + 2} \right) \qquad (11.5)$$

The gradient force is proportional to the intensity gradient and points up the gradient when $m > 1$. The large gradient force can be obtained by the large depth of the laser beam, in which the stable trapping force requires the gradient force in the $-\hat{z}$ direction against the direction of incident light (dark soliton valley) as shown in Figure 11.5. The gradient field strength can be increased by increasing the laser beam numerical aperture (NA) and decreasing focal spot size. In principle, the potential well is produced among the gaps by two forces to confine nanorobots.

The proposed system demonstrates that the nanorobots are stably trapped inside the potential wells and transported by the optical tweezers. The orthogonal tweezer spins can be formed within the system and detected simultaneously at the output ports. Surgical nanorobots can be programmed or driven by a human surgeon and act as a semi-autonomous on-site surgeon inside the human body. A device could perform various functions such as searching for pathology and then diagnosing and correcting lesions by nanomanipulation, coordinated by an onboard computer. A nanorobot transportation system has been designed to improve the surgical, pharmacological, and therapeutic properties of nanorobots. The strength of this system is their ability with targeted site feeding nanorobots, which can be used for surgical or drug delivery through cell membranes and into cell cytoplasm. On-site efficiency is important because many diseases depend upon processes within the cell and can only be impeded by drugs that make their way into the cell.

The important system characteristic of required dynamic behavior of the tweezers can be obtained by tuning some parameters of the system including control port input signal, coupling coefficient, and ring radius. In this adaptation, it is modified by using a liquid cladding inserted to cover an InGaAsP/InP waveguide core as shown in Figure 11.1; the proposed system consists of two states. First, a device called

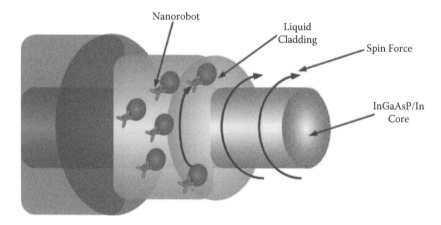

FIGURE 11.6   Nanorobot transportation system.

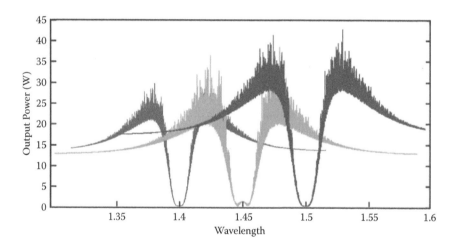

FIGURE 11.7   Optical tweezers (array) for many molecules trapping.

a *PANDA ring resonator* with gold coupling is used to generate the spin potential well for nanorobot trapping [25]. Second, the trapped nanorobots can be moved and rotated to the required destination by the generated potential wells. Figure 11.6 shows nanorobot transportation, in which the microgel/nanogel is used in the waveguide all the way through the neuronal cells. The output of optical tweezers from throughput and drop ports with different ring radii is simulated and shown in Figure 11.7. The results obtained have shown that the optical tweezers can trap, move, and rotate nanorobots to the required destination. In application, the trapped nanorobots can be moved and rotated to the required destinations, which can be useful for many applications, especially in targeted medicine, surgery, diagnosis of tissue tumors, drug delivery, and therapy.

11.3 CONCLUSION

A nanorobot transportation system using optical tweezers within a modified optical add–drop filter known as a PANDA ring resonator with gold coupling is proposed. A potential well is generated by spin within the system and detected simultaneously at the output ports by using dark and bright solitons. The orthogonal tweezers can be formed by dark and bright soliton pairs corresponding to the left-hand and right-hand rotating solitons under the resonant condition. In the system, liquid cladding is inserted to cover an InGaAsP/InP applied waveguide core, in which the nanorobot is trapped by optical spin tweezer. The result demonstrated that optical spin tweezers can be generated by the proposed system and the trapped nanorobots can be moved and rotated to the required destination for surgery and treatment. Furthermore, an array of molecular spins, trapped nanorobots, can be generated and detected by using the proposed system. In application, the proposed system can be useful for medical applications such as human surgeon diagnosis, cancer surgery, drug delivery treatment, and postsurgery therapeutic applications.

REFERENCES

1. G. D. M. Jeffries, J. S. Edgar, Y. Zhao, J. P. Shelby, C. Fong, and D. T. Chiu, Using polarization-shaped optical vortex traps for single-cell nanosurgery, *Nano Lett.*, 7 (3) 415–420, 2007.
2. A. Cavalcanti, B. Shirinzadeh, R. A. Freitas, Jr., and T. Hogg, Nanorobot architecture for medical target identification, *Nanotechnology*, 19, 015103, 2008.
3. S. M. Douglas, I. Bachelet, and G. M. Church, DNA robot could kill cancer cells, *Science*, 335, 831–834, 2012.
4. S. M. M. R. Al-Arif, N. Quader, A. M. Shaon, and K. K. Islam, Sensor based autonomous medical nanorobots: A cure to demyelination, *Multi J. Sci. Technol.* (JSAN), 2 (11) 1–7, 2011.
5. A. Cavalcanti, B. Shirinzadeh, and L. C. Kretly, Medical nanorobotics for diabetes control, nanomedicine: Nanotechnology, biology and medicine, *Patient Research*, 4 (2) 127–138, 2008.
6. S. X. Guo, F. Bourgeois, T. Chokshi, N. J. Durr, M. A. Hilliard, N. Chronis, and A. Ben-Yakar, Femtosecond laser nanoaxotomy lab-on-a chip for *in vivo* nerve regeneration studies, *Nat. Methods*, 5, 531–533, 2008.
7. A. Cavalcanti, B. Shirinzadeh, R. A. Freitas, Jr., and L. C. Kretly, Medical nanorobot architecture based on nanobioelectronics, *Recent Pat. Nanotechnol.*, 1 (1) 1–10, 2007.
8. N. N. Sharma and R. K. Mittal, Nanorobot movement: Challenges and biologically inspired solutions, *Int. J. Smart Sensing and Intelligent Systems*, 1 (1) 87–109, 2008.
9. T. Fukuda, M. Nakajima, P. Liu, H. El Shimy, Nanofabrication, nanoinstrumentation and nanoassembly by nanorobotic manipulation, *Int. J. Rob. Res.*, 28 (4) 537–547, 2009.
10. J. Fu and H. Yan, Controlled drug release by a nanorobot, *Nat. Biotechnol.*, 30, 407–408, 2012.
11. J. Elbaz and I. Willner, DNA origami: Nanorobots grab cellular control, *Nat. Mater.*, 11, 276–277, 2012.
12. R. K. Soong, G. D. Bachand, H. P. Neves, A. G. Olkhovets, H. G. Craighead, and C. D. Montemagno, Powering an inorganic nanodevice with a biomolecular motor, *Science*, 290 (5496) 1555–1558, 2000.
13. S. M. Douglas, I. Bachelet, and G. M. Church, A logic-gated nanorobot for targeted transport of molecular payloads, *Science*, 335 (6070) 831–834, 2012.

14. A. Ashkin, J. M. Dziedzic, J. E. Bjorkholm, and S. Chu, Observation of a single-beam gradient force optical trap for dielectric particles, *Opt. Lett.*, 11, 288–290, 1986.

15. K. C. Neuman and A. Nagy, Single-molecule force spectroscopy: Optical tweezers, magnetic tweezers and atomic force microscopy, *Nat. Methods*, 5, 491–505, 2008.

16. L. Huang, S. J. Maerkl, and O. J. F. Martin, Integration of plasmonic trapping in a microfluidic environment, *Opt. Express*, 17, 6018–6024, 2009.

17. K. C. Neuman and S. M. Block, *Optical trapping, Rev. Sci. Instrum.*, 75, 2787–2809, 2004.

18. L. J. Mathieu, R. Maurizio, and Q. Romain, Plasmon nano-optical tweezers, *Nat. Photonics.*, 5, 349–356, 2011.

19. M. A. Jalil, M. Tasakorn, N. Suwanpayak, J. Ali, and P. P. Yupapin, Nanoscopic volume trapping and transportation using a PANDA ring resonator for drug delivery, *IEEE Trans. Nanobioscience*, 10, 106–112, 2011.

20. M. S. Aziz, N. Suwanpayak, M. A. Jalil, R. Jomtarak, T. Saktioto, J. Ali, and P. P. Yupapin, Gold nanoparticle trapping and delivery for therapeutic applications, *Int. J. Nanomed.*, 7, 11–17, 2012.

21. K. Sarapat, N. Sangwara, K. Srinuanjan, and P. P. Yupapin, Novel dark–bright optical solitons conversion system and power amplification, *Opt. Engineering*, 48, 045004–045007, 2009.

22. A. Ashkin, Forces of a single-beam gradient force on a dielectric sphere in the ray of optic regime, *Biophys. J*, 61, 569–582, 1992.

23. I. H. Deutsch, P. M. Alsing, J. Grondalski, S. Ghose, P. S. Jessen, and D. L. Haycock, Quantum transport in magneto-optical double-potential wells, *J. Opt. B: Quant. and Semiclass. Opt.*, 2, 633–644, 2000.

24. K. Svoboda and S. M. Block, Biological applications of optical forces, *Annu. Rev. Biophys. Biomol. Struct.*, 23, 247–283, 1994.

25. N. Thammawongsa, N. Moongfangklang, S. Mitatha, and P. P. Yupapin, Novel nano-antenna system design using photonic spin in a PANDA ring resonator, *PIER Lett*, 31, 75–87, 2012.

12 Radiotherapy Using Nano-Antennas

12.1 INTRODUCTION

The use of radio frequency for medical application is increasing rapidly, especially for nanomedicine, which shows the most imperative role of human daily life in diagnostic and therapeutic aspects. Generally, radio frequency for therapeutic use known as *radiotherapy* involves the use of high-energy radiation from X-rays, γ-rays, neutrons, electrons, protons, and other sources to kill cancer cells and shrink tumors. There are two main types of radiotherapy: external-beam radiation therapy and internal radiation therapy (brachytherapy). The external-beam radiation therapy (EBRT) directs radiation beams at the tumor from outside the patient's body and normally the radiation beam is generated by a linear accelerator whereas the internal beam radiation therapy is the therapy in which a single or multiple radioactive sources are placed inside the patient's body [1]. An antenna is an essential component for signal radiation, which changes radiation intensity, or to exploit the transmitted power [2]. The surface plasmon resonance [3,4] characteristics describe the optical frequencies and exploit them to balance the drawbacks of antenna systems in the very-high-frequency spectrum range. Gold nanoparticles are extensively used as a nano-antenna due to perfect conduction and controllable size distribution, long-term stability, high homogeneity, and they also contribute plasmon resonances in the visible spectrum [5,6,7]. There are various types of nano-antennas available for therapeutic use, for instance, nanospheres and nanorods [5,8], bowtie nano-antennas [9], Yagi–Uda nano-antennas [10], and dipole nano-antennas [11].

From the recent progress in resonant subwavelength optical nano-antenna, researchers are offered a continuum of electromagnetic spectrum from radio frequencies all the way up to X-rays. There are enormous nano-antenna research applications such as nanobiotechnology [12], sensing [13], imaging [14], energy harvesting [15], and therapeutic use [16]. The nano-antenna for radiotherapeutic use has been extensively investigated, for instance, the use of fluorescence or the light emitter property of nano-antenna can be contributed to cancer cell detection and imaging, which takes advantage of the diagnostic imaging techniques and introduces probes to measure the expression of indicative cancer cell markers at different stages of diseases. In the case of diagnostic purposes, the surface plasmon resonance phenomenon include Mie scattering, surface plasmon absorption, surface-enhanced luminescence, and surface-enhanced Raman scattering (SERS) of gold nanoparticles, which has been used for selective therapeutic methods for cancer. Radiotherapy is the use of ionizing radiation for the curative, palliative, and prophylactic treatment of almost every type of solid cancer. In practice, thermotherapy [17,18] is a method of utilizing hyperthermia directed toward the body tissues for the purpose of damaging protein and structures within cancerous cells. For instance, the radio frequency at 13.6 MHz is used to heat 5 nm gold nanoparticles by generating high E-field strength. Consequently, it produces temperature at about 50.6±0.2°C for cancer cell killing [19]. Besides this research, three types of graphitic shelled-magnetic core nanoparticles (C-Fe, C-Fe/Co, and C-Co) are induced to generate localized heating by radio frequency at 350 kHz, in which it is experimentally confirmed that C-Fe nanoparticles have a high ability to kill cancer cells to over 99% [20]. In addition, the effective strategy for tumor cell killing is a combination of gold nanoparticles with a diameter of 1.9 nm by intravenous injection into mammary tumor-bearing mice and 250 kVp of X-rays, in which the one-year survival rate of mice treatment was 86% versus 20% for X-ray therapy alone and dropping to 0% for gold nanoparticle therapy alone [21]. Furthermore, nano-antenna is used to support advanced drug targeting and delivery as illustrated [18].

In operation, nanoradio power transmission is proportional to the input power, which is limited by the available power of the laser diodes. Recently, the use of photonic (optical) spins has been investigated for various applications [17,18] such as long distance optical transport, communication network and security, quantum computer, and communication. Besides the spin mechanism, the nano-antenna system design using photonic spin in a PANDA ring has been proposed by Thammawongsa and his colleagues [24]. In the system, the transverse electric (TE) and transverse magnetic (TM) fields are generated within the PANDA ring resonator by a soliton pulse, while the spin states are induced via an aluminum plate coupled to the microring resonator. The merit of such a system is that it can be used to generate the THz frequency for many aspects of applications. However, there is much research on radiotherapy but there is little information available on nano-antenna for therapeutic use. Therefore, this chapter proposes a new system of radio frequency therapy, which cooperates nano-antenna design with optical spin manipulation for treatment of a variety of malignancies. This is a simulation work, in which the advantage of optical spin is that there is no interference effect of light during propagation within the system, whereas finally the required transverse electric and magnetic field strengths can be controlled and obtained, which is of benefit to the hyperthermia system. In application, the proposed system can be used incorporating diagnostic/screening ability, stereotactic radiosurgery/surgical technique, and concurrent chemotherapy and system therapies.

12.2   RADIOTHERAPY USING NANO-ANTENNAS

Optical dipole is formed by spin manipulations [22,23], which are used for the conversion of dark–bright soliton pulse. The system using a PANDA ring resonator formed the orthogonal set of dark–bright soliton pairs, which can be decomposed into right and left circularly polarized waves. The relative phase of the two output light signals after coupling into the optical coupler is $\pi/2$. This means that the signals coupled into the drop port (Dr) and the through port (Th) have acquired a phase of π with respect to the input port (In) signal. The input and control fields at the input port and control port (Ct) are formed by the dark and bright optical solitons as described by Muhammad et al. [22] and Glomglome et al. [23].

Nano-antenna differs from radio frequency antenna in two important respects: high losses at optical frequencies, in which the assumption of a perfect electrical conductor is no longer valid (surface plasmon polarization). The most important different response of these structures is the subwavelength field confinement. Therefore, serious efforts are being devoted to extend current understanding from radio frequency antennas to nano-antenna counterparts. The model of an infinitesimal dipole is used to discuss the different character of the fields in near-field and far-field regions. Such models further help to appreciate the difference between the radiated power and the power stored in the near field of an antenna. The nano-antenna system design is shown in Figure 12.1. It consists of a four transmitting, four nanorod parts (dipole) connected through an optical channel. Electrons flow into the channel through

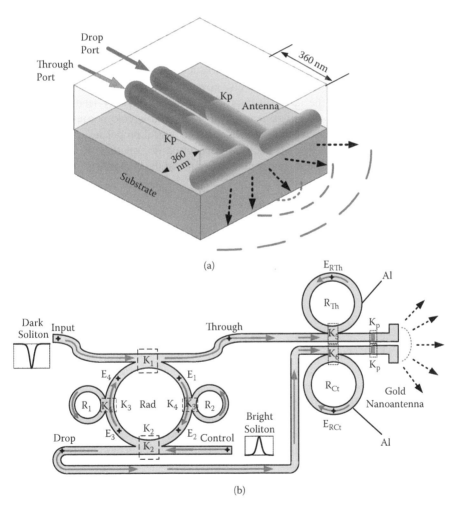

FIGURE 12.1 (a) Schematic of optical nano-antenna by optical spin manipulation generated within a PANDA ring resonator. (b) Nano-antenna using optical spin design.

the through port nanorod and drop port nanorod, which is coupled to get spin polarized when they reach channel interface. The spin properties with surface plasmon resonance can be employed to provide photothermal and radioactive energy to diseased tissues.

In the proposed system, the current density, dipole radiation, and field enhancement behavior are studied. The sinusoidal current distribution (I_s) is given in Equation (12.1) [25], I_m is the current amplitude, L is the total antenna length ($\lambda/2$), a is the radius of the wire (1/500), and k is the wave vector.

$$I_s(z) = I_m \sin\left(k\left(\frac{1}{2}L - |z|\right)\right) \tag{12.1}$$

To analyze the dipole antenna in more detail, a realistic current density of a dipole antenna is calculated using Pocklington's integral equation [26]. The calculation of the complex one-dimensional current density is performed by using the MATLAB function [27]. The antenna input impedance is defined [28] by the simplify impedance of the antenna to transmit and receive power at different lengths.

The optical spin is the factor to improve the optical intensity feeding into the nano-antenna. The most popular nano-antenna material is gold because it gives a good resonance and dielectric constant values. The optical properties of gold follow a Drude model. In this chapter, data has been fitted in Equation (12.2) [29]. The dispersion relation of the surface plasmon resonance at the interface is given by [30]:

$$\varepsilon_m(\omega) = \varepsilon_1 + i\varepsilon_2 \tag{12.2}$$

$$\varepsilon_1(\omega) = 1 - \frac{\omega_P^2}{\omega^2 + \Gamma^2} \tag{12.3}$$

$$\varepsilon_2(\omega) = \frac{\Gamma\omega_P^2}{\omega\left(\omega^2 + \Gamma^2\right)} \tag{12.4}$$

where ω_p and Γ denote the size and temperature-dependent plasma and collision frequencies of the nanoparticles, respectively, that is, $\omega_p = 1.36 \times 10^{16}$ rad s^{-1}, and $\Gamma = 1.05 \times 10^{14}$ rad s^{-1}.

The surface plasmon resonance intensity enhances the absorption of gold nano-antenna and increases the efficiency of the photothermal conversion process. Thus, the proposed system generates laser energy required to raise the temperature above the threshold. The extinction cross-section of a metal nanoparticle is derived by Mie theory and is given by

$$E(\omega) = \frac{24\pi R\varepsilon_m^{3/2}}{\lambda} \frac{\varepsilon_2(\omega)}{\left(\varepsilon_2(\omega) + 2\varepsilon_m\right)^2 + \varepsilon_2^2(\omega)} \tag{12.5}$$

A metal–dielectric interface supports charge–density oscillations along the interface and is called *surface plasma oscillations*. The surface plasmons are accompanied by a longitudinal electric field, which decays exponentially in metal as well as in dielectric medium. The electric field is maximized at metal–dielectric interface. The TM-polarization and exponential decay of an electric field can be measured by solving the Maxwell equation for semi-infinite media of metal and dielectric with an interface of metal–dielectric. The propagation constant (K_p) of the surface plasmon wave propagating along the metal–dielectric interface is given by

$$\kappa_p = \kappa_0 \sqrt{\left(\frac{\varepsilon_d \varepsilon_m(\omega)}{\varepsilon_d + \varepsilon_m(\omega)}\right)} \tag{12.6}$$

where κ_0 is the wave vector in air, ε_d is the relative permittivity of the dielectric, and $\varepsilon_m(\omega)$ is the dispersive relative permittivity of the metal. The total complex average power delivered to the antenna is demonstrated in Equation (12.4), which analyzed the power radiated in an infinitesimal dipole.

$$P = \frac{1}{2}|I_g|^2 Z_a = \frac{|V_g^2|}{2} \frac{R_a + X_a}{(R_a + R_g)^2 + (X_a + X_g)^2} = P_r + iP_{reac} \tag{12.7}$$

From the above equation, the imaginary part of P can be assigned to the reactive power P_{reac}, stored in the reactive near field. Here, P_{reac} represents the total reactive power. The real part of P is called the *radiate power P_r*, and Z_a is the antenna impedance, I_g and V_g are the source current and voltage, respectively.

The optical heating ΔT_{NP} of the nanorod model for a gold dipole antenna is given by Coulomb potential outside the particle as follows [36]:

$$\Delta T_{NP} = \frac{P}{4\pi R_{eq}\kappa_0} = \frac{E(\omega)I_r}{4\pi R_{eq}\kappa_0} \tag{12.8}$$

where the radius of a nanorod model is $R_{eq} = [(3D-d)d^2/16]^{1/3}$ and the irradiance of incoming spin light is I_r (this fixed by experimental setup). The irreversible temperature damage to diseased cells occurs at the temperature within the range of 40°C to about 80°C by varying the power input, antenna length, and wavelength for destruction of various targeted tumor cells. Moreover, the proposed system is able to sense tumors for diagnosis and therapy in a single network system for different frequency, noninterference, and various targeted tumors.

12.3   RESULTS AND DISCUSSION

In operation, the orthogonal soliton sets can be generated by using the PANDA ring resonator system. The optical field is fed into the ring resonator system, where $R_1 = R_2 = 2.5$ μm, $R_{ad} = 30$ μm, and $R_{Th} = R_{Ct} = 20$ μm. To form the initial spin states, the magnetic field is induced by an aluminum plate coupled on AlGaAs waveguides. In this simulation, the coupling coefficient ratios $\kappa_1{:}\kappa_2$ are 50:50, 90:10, 10:90 and the ring radii $R_{ad} = 30$ μm, $A_{eff} = 0.25$ μm², $n_{eff} = 3.14$ (for InGaAsP/InP) [23], $\alpha = 0.1$ dB/mm, $\gamma = 0.01$, and $\lambda_0 = 1.55$ μm as shown in Figure 12.1. The optoelectronic fields generated by dark solitons are pumped based on through port and drop port microring resonators at center wavelength 1.45 μm. Many soliton spins are detected at through (spin-up) and drop (spin-down) ports of a PANDA ring resonator as shown in Figure 12.2. The optoelectronic spin manipulation generated within a PANDA ring resonator is shown in Figure 12.3, where the output signals are randomly obtained at the through and drop ports [23], where the random transverse electric (TE) and transverse magnetic (TM) fields of the solitons corresponding to the left-hand and right-hand photons can be generated and detected. The angular momentum of either $+\hbar$ or $-\hbar$ is imparted to the object when a photon is absorbed by an object, where two possible spin states known as optoelectronic spins are exhibited.

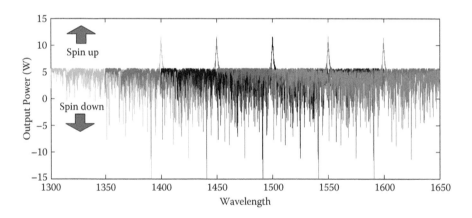

FIGURE 12.2 Many soliton spins obtained at through (spin-up) and drop (spin-down) ports.

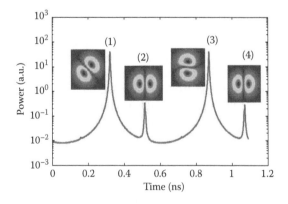

FIGURE 12.3 Spin rotations with time in a PANDA ring resonator.

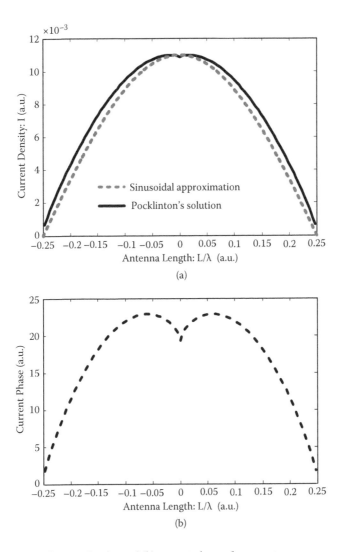

FIGURE 12.4 (a) Current density and (b) current phase of nano-antenna.

The proposed nano-antenna system can be divided into three parts. First, the highest current density and phase are shown in Figure 12.4. Second, input impedance for a design dipole nano-antenna with a length-to-radius ratio L/a is shown in Figure 12.5. And finally, the optical high power source [31] initiates the highest surface plasmon resonance, which is vital to generate optical intensity and polarization. In this proposed work, the power radiation and field enhancement are illustrated in Figure 12.6, in which the power transmission from the antenna and normalized field enhancement are compared by using the wavelength and frequency domains. This result presents the optical spin generation method, in which the radio frequency can be radiated by using the simple nano-antenna model.

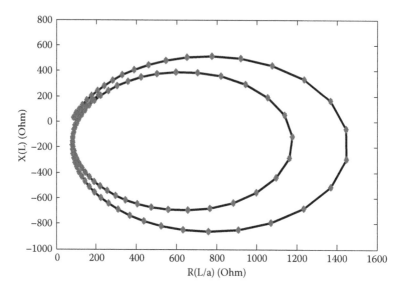

FIGURE 12.5 Impedance of nano-antenna (L/a).

Our proposed therapeutic system is composed of five parts (Figure 12.7a–e). Figure 12.7a is the input signal part, where the dark and bright solitons are fed into a device called a *PANDA ring resonator*, the ring radius is $R_1 = R_2 = 2.5$ μm at the input port and control port, respectively. The coupling coefficient ratios $\kappa_1, \kappa_2, \kappa_3$, and κ_4 are 10:90, 10:90, 20:80, and 20:80. Two signals are modulated by cross-phase modulation in order to form the orthogonal set of dark–bright soliton pairs, which can be decomposed into right and left circularly polarized waves. The output intensities of spin-injected for transverse electric (TE) and transverse magnetic (TM) fields are generated at throughput and drop ports before delivery. The output intensities of spin-injected by TE and TM fields are generated by using a PANDA ring resonator. Normally, a bus network is the most commonly used in computer networks. However, recently there is some research that applies bus networks for therapeutic applications [32,33]. In order to form the initial spin states, the magnetic field is induced by an aluminum plate coupled on InGaAsP/InP waveguides, in which the optoelectronic spin-up and spin-down states are generated via the dual microring resonators in Figure 12.7b, whereas the ring radius are $R_{Th} = R_{Ct} = 20$ μm and $\kappa_5 = \kappa_6 = 20:80$. The optoelectronic fields generated by a dark–bright soliton are pumped via through and drop ports of a microring resonator at the center wavelength 1.45 μm. As a result, the microring resonator can perform phase shift of the drop and through port output, in which the ring radii can be tuned. After that, the optoelectronic spin state of dual microring resonators are fed into the nano-antenna in Figure 12.7c. In practice, the gold properties are appealed to establish the nano-antenna for biomedical applications. The gold particles are nontoxic to human cells [34] and have radioactive properties of AU-198 [35], which fulfills criteria for an ideal candidate for radiotherapeutic applications. The soliton spin at the drop and through port of the microring resonator distributes the electromagnetic field to the disease cells. The schematic

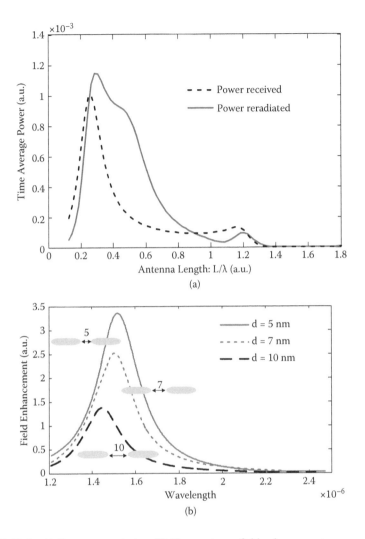

FIGURE 12.6   (a) Power transmission. (b) Nano-antenna field enhancement.

of optical nano-antenna generated by a PANDA ring resonator and parameters of gold nano-antenna are described in Figure 12.1. However, all parameters are chosen closely to the practical parameters, which means that the proposed device can be fabricated and realized [25,26]. In Figure 12.7d, the shaded area represents the diseased tumor cells (target cells), while the breast tumor area is as shown in Figure 12.7e. In therapy, for the most part, the attempt is to focus all radiation on the targeted cells and the tumor area. The adjacent normal tissues receive a little radiation, whereas the proposed system is the internal radiation therapy system in which a nanosystem is placed inside the body near the tumor area. In order to generate the hyperthermia at specific frequencies to heat the targeted cells, Rakic et al. [29] have already described the irreversible damage to diseased cells occurring

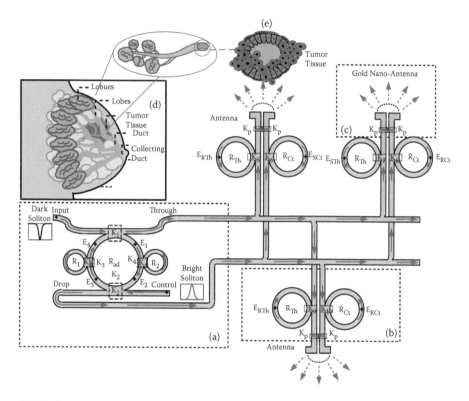

FIGURE 12.7 Nanoradiotherapeutic system. (a) Many spins generation. Parts (b), (c), (d), and (e) are tumor cells and tissue. The system can be fabricated and formed by a thin-film device. It is totally in the millimeter scale, which can be formed by a single chip (on-chip).

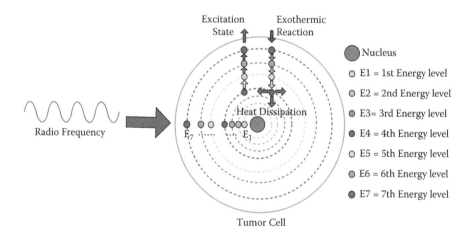

FIGURE 12.8 Model of wave interaction with biological tissue.

at a temperature within the range of 40°C to about 46°C. The model of wave interaction with biological tissue can be described by using the atomic Bohr theory as shown in Figure 12.8. When the radio frequency emits signals through the tumor cells, the orbital electrons can be excited and moved to the excited state, in which the direct or indirect ionizations can be generated. The energy exchange between atoms and molecules occurs, where finally, the system (tumor cells) needs to adjust itself to be in the equilibrium state. Consequently, the heat dissipation or energy exchange is released by the exothermic reaction process, where the tumor cells can be destroyed, that is, killed. Accordingly, the healthy cells are capable of surviving at temperatures up to around 46.5°C [18]. A gold nano-antenna is recommended to conform to such requirements, in which the required electromagnetic field strengths (near fields) can be generated, controlled, and radiated to the target cells, where finally, the required therapeutic cell treatments can be achieved.

12.4 CONCLUSION

This chapter proposed an interesting system that can be used for multipurpose therapeutic applications, especially for a variety of tumor malignancies. The powerful terahertz frequency source with different wavelengths can be used for suitable treatment of targeted cells, in fact, they are the orthogonal soliton pairs generated by the two optical dipole components (TE and TM waves) within a PANDA ring resonator. The optoelectronic spin-up and spin-down states formed at *Th* and *Dr* port can be used to form the radio frequency oscillation. The nano-antenna transmission is controlled by the ring parameters, where in this case, the suitable thermal distribution of a gold nano-antenna is 45°C, which is not harmful for healthy cells. The system can be fabricated in a form of thin film, which can be embedded near the tumor cells. In application, the designed system can be embedded within a human body and closed to the required therapeutic points (cells). In addition, the optical dipole high effective power nano-antenna resonance can be achieved, realized, and collaborated on in the diagnostic/screening ability, stereotactic radiosurgery/surgical technique, and concurrent chemotherapy and system therapies.

REFERENCES

1. B. J. Roxworthy, K. D. Ko, and A. Kumar et al., Application of plasmonic bowtie nano-antenna arrays for optical trapping, *Nano Lett.*, 12:796–801, 2012.
2. C. A. Balanis, *Antenna Theory: Analysis and Design*. 2nd ed., New York: John Wiley & Sons, Inc., 1996.
3. P. Cao, X. Zhang, W. J. Kong et al., Super resolution enhancement for the super lens with anti-reflection and phase control coatings via surface plasmons modes of asymmetric structure, *Prog. Electromagn Res. (PIER)*, 119, 191–206, 2011.
4. T. Suyama and Y. Okuno, Enhancement of TM-TE mode conversion caused by excitation of surface plasmons on a metal grating and its application for refractive index measurement, *Prog. Electromagn Res. (PIER),* 72, 91–103, 2007.
5. D. Pissuwan, S. M. Valenzuela, and M. B. Cortie, Prospects for gold nanorod particles in diagnostic and therapeutic applications, *Biotechnol Genet Eng (BGER)*, 25, 93–112, 2008.

6. A. Moscatelli, Gold nanoparticles afloat, *Nat. Mater.*, 11, 8, 2011.

7. G. H. Lin, R. Abdu, and J. O. M. Bockris, Investigation of resonance light absorption and rectification by subnanostructures, *J. Appl. Phy.*, 80, 565–568, 1996.

8. N. C. Dmitry, K. Christian, and V. Z. Sergei, Plasmonic nanoparticle monomers and dimers: From nano-antennas to chiral metamaterials, *Appl. Phys. B*, 105, 1–17, 2011.

9. H. Guo, T. P. Meyrath, T. Zentgraf et al., Optical resonances of bowtie slot antennas and their geometry and material dependence, *Opt. Express,* 16, 7756–7766, 2008.

10. T. H. Taminiau, F. D. Stefani, and N. F. van Hulst, Enhanced directional excitation and emission of single emitters by a nano-optical Yagi-Uda antenna, *Opt. Express.,* 16, 10858–10866, 2008.

11. K. Sendur and E. Baran, Near-field optical power transmission of dipole nano-antennas, *Appl. Phys. B: Lasers and Optics*, 96, 325–335, 2009.

12. K. P. Tallury, A. Malhotra, L. M. Byrne et al., Nanobioimaging and sensing of infectious diseases, *Adv. Drug Delivery Rev.,* 62, 424–437, 2010.

13. N. Liu, M. L. Tang, M. Hentschel et al., Nanoantenna-enhanced gas sensing in a single tailored nanofocus, *Nat. Mater.,* 10, 631–636, 2011.

14. G. R. L. Duc, I. Miladi, C. Alric et al., Toward an image-guided microbeam radiation therapy using gadolinium-based nanoparticles, *ACS Nano.,* 5, 9566–9574, 2011.

15. D. K. Kotter, S. D. Novack, W. D. Slafer et al., Theory and manufacturing processes of solar nano-antenna electromagnetic collectors, *J. Sol. Energy Eng.,* 132, 011014–011023, 2010.

16. W. Roa, X. Zhang, L. Guo et al., Gold nanoparticle sensitize radiotherapy of prostate cancer cells by regulation of the cell cycle, *Nanotechnology*, 20, 375101–375110, 2009.

17. P. P. Connell and S. Hellman, Advances in radiotherapy and implications for the next century: A historical perspective, *Cancer Res.,* 62, 383–392, 2009.

18. N. P. Praetorius and T. K. Mandal, Engineered nanoparticles in cancer therapy, *Recent Pat. Drug Deliv. Formul.,* 1, 37–51, 2007.

19. D. E., Kruse, D. N. Stephens, H. A. Lindfors et al. A radio-frequency coupling network for heating of citrate-coated gold nanoparticles for cancer therapy: Design and analysis, *IEEE Trans. Biomed. Eng.,* 58, 2002–2012, 2011.

20. Y. Xu, M. Mahmood, A. Fejleh et al., Carbon-covered magnetic nanomaterials and their application for the thermolysis of cancer cells, *Int. J. Nanomed.,* 5, 167–176, 2010.

21. A. Mesbahi, A review on gold nanoparticles radiosensitization effect in radiation therapy of cancer, *Rep. Practical Oncol. Radiother.,* 15, 176–180, 2010.

22. F. D. Muhammad, C. Teeka, J. Ali et al., Optical spin manipulation using dark–bright soliton behaviors in a PANDA ring resonator, *Microwave Opt Technol Lett.,* 54, 987–990, 2012.

23. S. Glomglome, I. Srithanachai, C. Teeka et al., Optical spin generated by a soliton pulse in an add-drop filter for optoelectronics and spintronics use, *Optics & Laser Technology,* 44, 1294–1297, 2012.

24. N. Thammawongsa, N. Moongfangklang, S. Mitatha et al., Novel nano-antenna system design using photonic spin in a PANDA ring resonator, *PIER*, 31, 75–87, 2012.

25. R. King and C. W. Harrison, The distribution of current along a symmetrical center-driven antenna, *Proc IRE.,* 31, 548–567, 1943.

26. H. C. Pocklington, Electrical oscillations in wires, *Proc. Cambridge Philosophical Society*, 9, 324–332, 1897.

27. P. Biagioni, J. S. Huang, and B. Hecht, Nanoantennas for visible and infrared radiation, *Reports on Prog. Phys.,* 75, 24402–24441, 2012.

28. C. A. Balanis, Antenna theory: A review, *Proc IEEE*, 80, 7–22, 1992.

29. A. D. Rakic, A. B. Djuriic, J. M. Elazar et al., Optical properties of metallic films for vertical-cavity optoelectronic devices, *Appl. Opt.*, 37, 5271–5283, 1998.

30. W. L. Barnes, A. Dereux, and T. W. Ebbesen, Surface plasmon sub-wavelength optics, *Nature*, 424, 824–830, 2003.
31. J. Alda, J. M. R. García, J. M. Alonso et al., Optical antennas for nano-photonic applications, *Nanotechnology*, 16, S230–S234, 2005.
32. S. Mitatha, N. Moogfangklang, M. A. Jalil et al., Proposal of Alzheimer's diagnosis using molecular buffer and bus network, *Int. J. Nanomed.*, 6, 1–8, 2011.
33. H. J. Xiaohua, K. J. Prashant, I. H. El-Sayed et al., Gold nanoparticles: Interesting optical properties and recent applications in cancer diagnostics and therapy, *Nanomedicine*, 2, 681–693, 2007.
34. M. A. Jalil, N. Moongfangklang, K. Innate et al., Molecular network topology and reliability for multipurpose diagnosis, *Int. J. Nanomed.*, 6, 2385–2392, 2011.
35. K. V. Katti, R. Kannan, K. Katti et al., Hybrid gold nanoparticles in molecular imaging and radiotherapy, *Czech J. Phys.*, 56, D23–D34, 2006.
36. B. Guillaume, Q. Romain, and F. J. García de Abajo, Nanoscale control of optical heating in complex plasmonic systems, *ACS Nano.*, 4, 709–716, 2010.

13 Neuron Cell Communications

13.1 INTRODUCTION

Communication among cells plays a crucial role in the human body mechanism by sending and receiving signals. It conveys information for sensation, feeling, emotional responses, thoughts, learning, memory, cause of mental disorders, and any other function of the human brain [1]. The nerve cell communication model is as shown in Figure 13.1, where the signals from the surrounding environment or other cells such as a response trigger signal must be communicated across a cell membrane. Signal information can cross the membrane passed through the movement of an electrical impulse and contact both outside and inside of the cell interacting with receptor proteins [2]. In this case, the correct cell receptors will respond to the signal on their surfaces [3]. Generally, a neuron or nerve cell is the key player in the nervous system activity. Within the neuron, two types of phenomena, chemical and electrical, are involved in the nerve impulse process. There are three main parts of a neuron. First, a dendrite is the thin fiber that extends for hundreds of micrometers in many branched tendrils that arise from the cell body, to receive information from other neurons. Second, soma or the cell body, is the majority of the neuron's basic cellular functioning. For instance, the soma of a neuron can vary from 4 to 100 micrometers in diameter. The last important one, a long thin fiber called an *axon*, transmits nerve impulses to other neurons [4]. Neurons present in many different shapes and sizes, which can be categorized into two types by their function and morphology. Type I with long axons are used to move signals over long distances; the basic morphology represented by spinal motor neurons, consists of a soma and a long thin axon covered by the myelin sheath. The end of the axon has branching terminals called *axon terminals,* which are used to release neurotransmitters into a gap called the *synaptic cleft* between the terminals and the dendrites of the next neuron (4 nm). The adult human brain is estimated to contain from 10^{14} to 5×10^{14} synapses. Around the soma there is a dendrite branching tree that receives signals from other neurons. Type II have short axons, which can often be confused with dendrites. The emergence of rhythmic structure contributes to different activity in the cell communication network [5,6].

After the discovery of neuronal inhibition in the early 1950s, it was accepted that information transmission via neurotransmitter molecules (chemical synapses) represents the major form of signaling among central nervous system (CNS) neurons. Recently, a wide variety of observations of cell communication was proposed [7,8], in which the developed electrical synapses were presented in a subset of neuronal connections. The electrical synapses are usually bidirectional transmission, ideal for

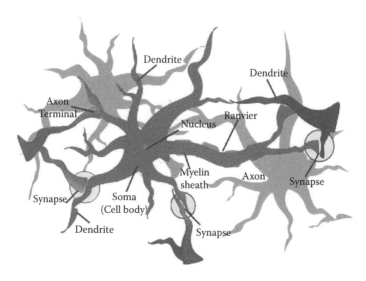

FIGURE 13.1 Nerve cell communication model.

synchronized behavior in a syncytium, very fast signal transmission, temperature insensitive, excitatory, and limited amplitude about 1 mV [9,10]. Investigations of the effects of neural activity on the strength of electrical synapses in the mammalian CNS were presented. Haas, Zavala, and Landisman [11] recorded the electrical activity from pairs of neurons in the rat thalamic reticular nucleus (TRN) that affects long-term depression (LTD) in the thalamus of the mammalian brain. All sensory messages are conveyed to the cortex via the thalamus. The bursts of spikes in pairs of coupled TRN neurons resulted in the strength of electrical synapses. The limitation of this investigation in interference to adjacent cells occurred when coupled between paired neural cells. The use of biosensors for diagnosis of diseases, food testing, and the environmental detection of biological agents has increased dramatically over the past few decades. Recently, a fiber-optic localized surface plasmon resonance (FO-LPR) chemical and biochemical sensing platform for label-free and real-time detection has been developed [12,13]. The unclad section of an optical fiber is modified with self-assembled gold nanoparticles which are functionalized with a receptor, where the characteristics of gold nanoparticles exhibit a strong absorption band that results when the incident photon frequency is resonant with the conduction electron oscillation and is known as the localized plasmon resonance (LPR). A targeted signal can be detected or transmitted by gold nanoparticles called *gold nano-antenna* [14].

A new research vision, optical spin, shows great promise nowadays, in which a spin mechanism using bright and dark soliton conversion behaviors is successfully demonstrated by Sarapat et al. and Glomglome et al. [15,16], which consists of a modified add-drop optical filter known as a PANDA ring resonator. Soliton spins can be decomposed into left and right circularly polarized waves called a *dark–bright soliton pair*. Additionally, the use of a nanoscale device is a very important component for a nano regime; one of the interesting aspects is a nano-antenna, which is a nanoscale device in which the surface plasmon resonance (SPR) theory for signal propagation and detection can be formed and used [17–19]. Moreover, there are many optical spin applications such as high-data-rate communications [20], monitoring and spectroscopy, medical imaging [21], security, material spectroscopy and sensing [22], and biology and medicine [23]. For the cell-biological scale, optical nano-antenna has dominated features due to its unique features for fabrication and characterization, which cover the potential advantages in the detection of light showing polarization, tunable, targeted cells, and rapid time response. Consequently, this chapter demonstrates a neural cell communication model for bio-cell coupling with optical nano-antenna based on a PANDA ring resonator to generate the optical spin manipulation. A schematic of a nerve communication system design called *Spinnet* will be illustrated in the next section. The results obtained have shown that the spin mechanism using bright and dark soliton conversion increases the electrical coupling efficiency by optical spin nano-antenna for nerve communication network. In applications, the link between nerves and cells may be useful and realized for brain learning and applying knowledge, neurological disorders, therapeutic disability, and rehabilitation cell disease applications.

13.2   NEURON CELL MODEL

The fundamental mechanism of signal transmission within neurons is based on potential differences between inner and outer sides of the cell. The membrane potential is produced by the irregular distribution of electrically charged particles, or ions, sodium (Na+), potassium (K+), chloride (Cl-), and calcium (Ca^2+) [24]. The voltage difference across the membrane is affected by redistribution of electric charge. Ions or neurotransmitters enter and exit the cell through specific protein channels in the cell's membrane. The cell's membrane potential controls the channel changing for "open" or "close" [25]. Consequently, an impulse travels to the neuron by decreasing exceeded voltage difference in a certain threshold called *depolarization*. It occurs when positive ions enter the neuron, making it more susceptible to fire an action potential. On the other hand, hyperpolarization makes it less susceptible to fire when negative ions enter the neuron. Each neuron receives depolarizing and hyperpolarizing currents from many neurons. The neuron fires an action potential when the depolarizing currents (positive ions) minus the hyperpolarizing current (negative ions) exceed minimum intensity (threshold) [26].

A comprehensive nerve linkage network connection is required to describe nerve cell communication, where the junction between the axon tips of the sending neuron and the dendrite or soma of the receiving neuron is called *synapses*. The transferred communication between the tiny gap [27,28] is called the synapses *gap* or *cleft*. At gap junctions, such cells approach within about 3.5 nm of each other. The typical and most frequent type of synapse is the one in which the axon of one neuron connects to a second neuron by usually making contact at one of its dendrites or the cell body. The general information flow direction is from the axon terminal to the target neuron; therefore, the axon terminal is called *presynaptic*

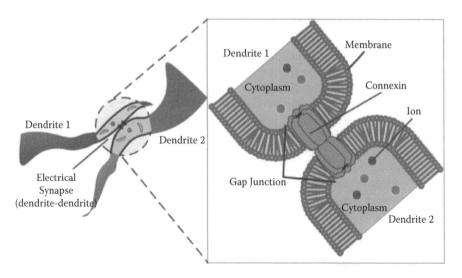

FIGURE 13.2   The electrical synchronized model by the electrical neurotransmitter.

(carries information toward a synapse) and the target neuron is called *postsynaptic* (carries information from a synapse). Most mammalian synapses are chemical; however, there is an uncomplicated electrical synapse form that allows directly transferred ionic current through one cell to the next. In the chemical synapses type, the incoming signal is transmitted when one neuron releases a neurotransmitter into the synaptic cleft, which is detected by the second neuron activate receptors opposite to the release site [29,30]. The synaptic delay for a chemical synapse is typically about 2 ms, while the synaptic delay for an electrical synapse may be about 0.2 ms. The electrical synapses take place in gap junctions as shown in Figure 13.2, in which the connexon channel in connexin [31] allows ions to pass directly from the cytoplasm of one cell to another one. The electrical synapses transmission is very rapid, thus, an action potential in the presynaptic neuron can be produced nearly instantaneously. The ventrobasal nucleus (VBN), the primary somatosensory area of the thalamus, was explained by Pozza et al. [32]. The electrical synapses were common between VBN neurons during the first postnatal week, and decreased during the second week as chemical synaptic circuits emerged. Parker et al. [33] calculated two isopotential neurons coupled directly by a single junction using values of the injected currents and voltage response of each cell junction conductance. Furthermore, a conductance-base neural model calculated the synchronizing action of electrical synapses between connexin [34]. There are several types of connexin in neurons such as Cx32, Cx36, Cx43, Cx45, and Cx47, where most of the electrical coupling between VBN cells required Cx36. The junction potential coupling from optical spin technique is analyzed and demonstrated by our proposed network.

The optical dipole interaction model in many substances illustrates the origin of various optical effects. At nanoscale, a photon from a source generator causes optical effect of light interaction in nanostructure. Polarization is a function of the nano-antenna cross-section absorption with the electromagnetic-induced dipole in atom and gradient intensity. One of the interesting schemes is that the optical dipole is formed by spin manipulations in a PANDA ring resonator [15], where a PANDA ring resonator is used for conversion of a dark–bright soliton pair. The orthogonal set of the dark–bright soliton pair is decomposed into right and left circularly polarized waves, where finally many spins can be generated and achieved; the theoretical review is given in the following details.

The relative phase of the two output light signals after coupling into the optical coupler is $\pi/2$. The signals coupled into the drop port and the through port obtained a phase of π with respect to the input port signal. The input and control fields at the input port and control port are produced by the dark and bright optical solitons as described in Sarapat et al. [15].

13.3 CELL COMMUNICATION MODEL

The Spinnet system using the PANDA ring resonator system to generate the orthogonal soliton sets is shown in Figure 13.3. In the simulation, the coupling coefficient ratios $\kappa_1{:}\kappa_2$ are 50:50, 90:10, 10:90, and $R_r = R_l = 5$ μm by using a microring, $R_{Th} = R_{add} = 15$ μm and the ring radii $R_{ad} = 25$ μm, $A_{eff} = 0.25$ μm², $n_{eff} = 3.14$ (for InGaAsP/InP)[16], $\alpha = 0.1$ dB/mm, $\gamma = 0.01$, $\lambda_0 = 1.55$ μm. The optical fields are generated by a dark–bright soliton pump based on through port

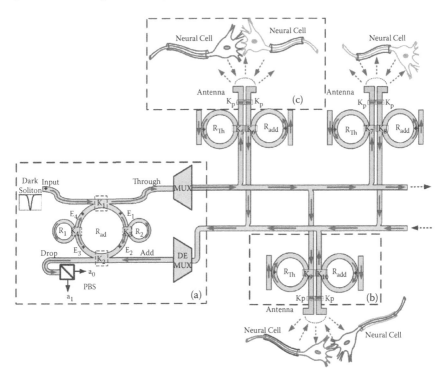

FIGURE 13.3 A schematic of nerve communication system.

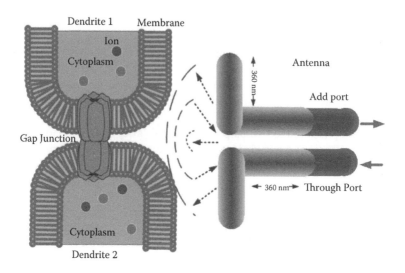

FIGURE 13.4   Cell coupling with nano-antenna model.

and drop port microring resonator at wavelength 1.55 μm. Many optical spins are detected at the through and add ports of a PANDA ring resonator. The soliton optical field is fed into the ring resonator system. The initial spin states form the magnetic field induced by a gold nano-antenna coupled on AlGaAs waveguides for optical spin-up and spin-down states. The high optical power source initiates the highest surface plasmon resonance, which is vital to generate optical intensity and polarization. The Spinnet nerve communication system design known as Spinnet is shown in Figure 13.4. All parameters were chosen closely to the practical parameter in which the proposed device can be fabricated and realized [17–19,35]. Part (a) is illustrated in the Figure 13.3, which added the multiplexer and demultiplexer for the transmission and detection difference wavelength in the network. In part (b), R_{Th} and R_{add} represents the add-drop to transmit and detect pulse signal. The coupling area is shown in part (c). Figure 13.5 shows the cell coupling between dendrite 1, dendrite 2, and nano-antenna. The optical nano-antenna is placed outside the neural cell near the gap junction about 5 nm for optimum radiation and detection. The adjacent normal nerve cells are able to receive the external radiations from the nano-antenna placed outside the neuron cell near the gap junction and adjacent normal cells can receive the signal. The coupling is established at specific frequencies to targeted cells.

As a result, the output signals are achieved at the through and drop ports as illustrated in Figure 13.5. Many potential wells can be generated by a PANDA ring and introduced by the nonlinear side ring effects, using the finite-difference time-domain (FDTD) method. The spontaneous activity occurred due to the interaction of the electrical fields between the gap junction and nano-antenna, which is induced by the wave-particle duality and energy discrepancy, in which the surface plasmon resonance of the device can be converted to the optical radiation via the nano-antenna; more details are found in Thammawongsa et al. [35]. The random

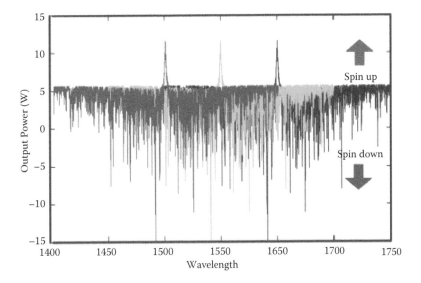

FIGURE 13.5 Optical spin-up and spin-down states in different wavelength.

transverse electric (TE) and transverse magnetic (TM) fields of the solitons corresponding to the left-hand and right-hand photons can be generated and detected. Furthermore, the angular momentum of either $+\hbar$ or $-\hbar$ is imparted to the object when a photon is absorbed by an object. There will be two possible spin states known as optical spins [16]. The extensions of electrical coupling have been observed experimentally [35,36] for impulse potential generation. As an approximation of the coupling strength, the hyperpolarizing current pulses (600 ms) are fed into first cell (*V1*), and coupled to neighbor cells (*V2*). A coupling coefficient is calculated as *V2/V1*. The optical dipole pairs were considered to be electrically coupled, which was the minimum level that could be reliably detected under the conditions of experiments in Parker, Cruikshank, and Connors [33]. Also, the junction conductance is estimated by using the values of the injected voltage and currents. The spontaneous firing of electrically coupled and spike transmission through electrical synapses is shown in Figure 13.6. The bursting driven is obtained by a simultaneous current, which is injected into both cells of coupled pairs in scale bars as 20 mV and 50 ms, where each membrane excitation patch has two important potential levels, the resting potential, which is the membrane potential value is preserved as long as nothing perturbs the cell is around −70 mV, and a higher value called the *threshold potential* is around −55 mV [37]. The electrical synaptic inputs cause the membrane to depolarize or hyperpolarize, which is the signal potential formed by the optical spin dark–bright soliton, which can cause the membrane potential to rise up or fall down.

The interaction of the electrical fields between the gap junction and nano-antenna can be described by using the atomic Bohr theory as the following details. When the radio frequency emits signal (energy) through neurons, the orbital electrons can be excited and moved to the excited state, in which the direct

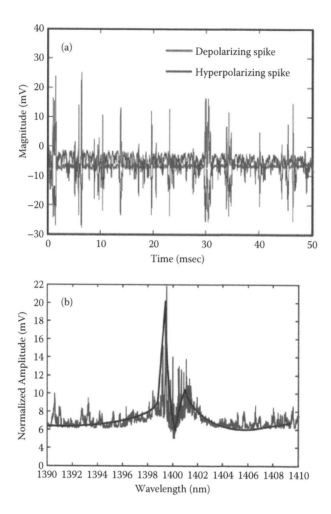

FIGURE 13.6 (a) Spontaneous firing of electrically coupled signal and (b) spike transmission through electrical synapses.

or indirect ionizations can be generated and energy exchanged. The energy exchange between electrons and neurons occurs, where finally, the system needs to adjust itself to be in the equilibrium state. Consequently, the energy exchange is released and the spontaneous emission occurs, which can be modeled by using the resistor-capacitor model to describe the gap junction's behavior as shown in Figures 13.7 and 13.8. In Figure 13.6b, by using the low pass filter for a synaptic equivalent circuit as shown in Figure 13.7, the presynaptic gap is connected by the postsynaptic conductor (C), which is parallel connected in the circuit. The result of resting potentials is shown in Figure 13.8, in which the system model is formed by the synaptic interconnection. The advantage of the proposed system is a simple system that can be fabricated to a nanodevice for a nerve communication system [38].

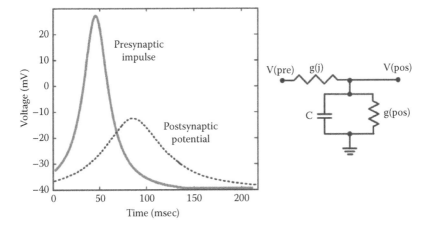

FIGURE 13.7 Gap junction of the postsynaptic capacitance behaves as low pass filters.

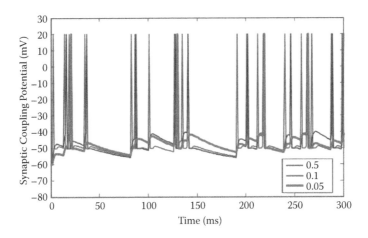

FIGURE 13.8 The steady-state responsive pattern of excitatory neurons for three different levels of the resting potentials, where the thickness is labeled as 0.5, 0.1, and 0.05.

13.4 CONCLUSION

A nanoscale communication network using ring resonators has been proposed for nerve cell communications. The optical dipole is generated and detected by using the metal coating waveguide, which is coupled to the nearby nerves or cells, which can be formed by the near-field effects. The coupling between optical dipole and nerve signals is formed and propagated with the optical networks at specific wavelength for targeted cells. Simulation results have shown that the use of optical dipoles in the optical networks can be formed, in which the dark and bright soliton pair can propagate for long distance in the network without electromagnetic interference. Optical spins and dipoles are established by the coupling effects between soliton pair and metallic waveguide, in which the TE and TM fields can be generated and detected.

The coupling signals between cells occur and demonstrate the spontaneous firing of electrically coupled and spike transmission through electrical synapses, where the normalized amplitude of about 20 mV is seen. The nerve signals in optical network can come across the dead region, which can reach the targeted cells and noninterference to adjacent cells. In applications, this network can be used in medical diagnosis and therapeutic application such as disability and rehabilitation.

REFERENCES

1. K. Meyl, Task of the introns, cell communication explained by field physics, *J. Cell Commun. Signal.*, 6, 53–58, 2012.
2. S. C. Lee, S. J. Cruikshank, and B. W. Connors, Electrical and chemical synapses between relay neurons in developing thalamus, *J Physiol,* 588, 2403–2415, 2010.
3. R. G. Kavasseri and R. Nagarajan, Synchronization in electrically coupled neural networks, *Complex Systems,* 16, 369–380, 2006.
4. M. V. Bennett and R. S. Zukin, Electrical coupling and neuronal synchronization in the mammalian brain, *Neuron,* 41, 495–511, 2004.
5. C. E. Floresa, S. Nannapanenia, K. G. Davidson, T. Yasumura, M. V. Bennett, J. E. Rash, and A. E. Pereda, Trafficking of gap junction channels at a vertebrate electrical synapse *in vivo, PNAS,* 109, E573-E582, 2012.
6. B. W. Connors and M. A. Long, Electrical synapses in the mammalian brain, *Annu. Rev. Neurosci.,* 27, 393–418, 2004.
7. B. M. Adhikari, A. Prasad, and M. Dhamala, Time-delay-induced phase-transition to synchrony in coupled bursting neurons, *Chaos,* 21, 0231161–0231167, 2011.
8. I. F. Akyildiz and J. M. Jornet, Electromagnetic wireless nanosensor networks, *Nano Commun. Netw.,* 1, 3–19, 2010.
9. J. R. Gibson, M. Beierlein, and B. W. Connors, Functional properties of electrical synapses between inhibitory interneurons of neocortical layer 4, *J. Neurophysiol.,* 93, 467–480, 2005.
10. K. L. Blethyn, S. W. Hughes, and V. Crunelli, Evidence for electrical synapses between neurons of the nucleus reticularis thalami in the adult brain in vitro, *Thalamus Relat Syst.,* 4, 13–20, 2008.
11. J. S. Haas, B. Zavala, and C. E. Landisman, Activity-dependent long-term depression of electrical synapses, *Science,* 334, 389–393, 2011.
12. Q. Zhang, C. Xue, Y. Yuan, J. Lee, D. Sun, and J. Xiong, Fiber surface modification technology for fiber-optic localized surface plasmon resonance biosensors, *Sensors,* 12, 2729–2741, 2012.
13. Y. Tian, W. Wang, N. Wu, X. Zou, and X. Wang, Tapered optical fiber sensor for label-free detection of biomolecules, *Sensors,* 11, 3780–3790, 2011.
14. B. Y. Hsieh, Y. F. Chang, M. Y. Ng, W. C. Liu, C. H. Lin, H. T. Wu, and C. Chou, Localized surface plasmon coupled fluorescence fiber-optic biosensor with gold nanoparticles, *Anal. Chem.,* 79, 3487–3493, 2007.
15. K. Sarapat, N. Sangwara, K. Srinuanjan, P. P. Yupapin, and N. Pornsuwancharoen, Novel dark-bright optical solitons conversion system and power amplification, *Opt. Eng.,* 48, 045004–045007, 2009.
16. S. Glomglome, I. Srithanachai, C. Teeka, S. Mitatha, S. Niemcharoen, and P. P. Yupapin, Optical spin generated by a soliton pulse in an add-drop filter for optoelectronic and spintronic use, *Opt. Laser Technol.,* 44, 1294–1297, 2012.
17. T. J. Seok, A. Jamshidi, M. Kim, S. Dhuey, A. Lakhani, H. Choo, P. J. Schuck, S. Cabrini, A. M. Schwartzberg, J. Bokor, E. Yablonovitch, and M. C. Wu, Radiation engineering of optical antennas for maximum field enhancement, *Nano Lett.,* 11, 2606–2610, 2011.

18. G. Armelles, J. B. González-Díaz, A. García-Martín, J. Miguel, G. Martín, A. Cebollada, M. U. González, S. Acimovic, J. Cesario, R. Quidant, and G. Badenes, Localized surface plasmon resonance effects on the magneto-optical activity of continuous Au/Co/Au trilayers, *Opt. Express*, 16, 16104–16112, 2008.

19. H. Fischer and O. J. Martin, Engineering the optical response of plasmonic nanoantennas, *Opt. Express*, 16, 9144–9154, 2008.

20. A. D. Rakic, A. B. Djuriic, J. M. Elazar, and M. L. Majewski, Optical properties of metallic films for vertical-cavity optoelectronic devices, *Appl. Opt.*, 37, 5271–5283, 1998.

21. S. Mitatha, N. Moogfangklang, M. A. Jalil, N. Suwanpayak, T. Saktioto, J. Ali, and P. P. Yupapin, Proposal of Alzheimer's diagnosis using molecular buffer and bus network, *Int. J. Nanomed.*, 6, 1–8, 2011.

22. K. V. Katti, R. Kannan, K. Katti, V. Kattumori, R. Pan Drapraganda, V. Rahing, C. Cutler, E. J. Boote, S. W. Casteel, C. J. Smith, J. D. Robertson, and S. S. Jurrison, Hybrid gold nanoparticles in molecular imaging and radiotherapy, *Czech J. Phys.*, 56, D23–D34, 2006.

23. M. S. Aziz, N. Suwanpayak, M. A. Jalil, R. Jomtarak, T. Saktioto, J. Ali, and P. P. Yupapin, Gold nanoparticle trapping and delivery for therapeutic applications, *Int. J. Nanomed.*, 7, 11–17, 2012.

24. Z. Li, T. Cai, J. Tian, J. X. Xie, X. Zhao, L. Liu, J. I. Shapiro, and Z. Xie, NaKtide, a Na/K-ATPase-derived peptide Src inhibitor, antagonizes ouabain-activated signal transduction in cultured cells, *J Biol. Chem.*, 284, 21066–21076, 2009.

25. A. G. Blankenship and M. B. Feller, Mechanisms underlying spontaneous patterned activity in developing neural circuits, *Nat. Rev. Neurosci.*, 11, 18–29, 2009.

26. K. Vervaeke, A. Lőrincz, P. Gleeson, M. Farinella, Z. Nusser, and R. A. Silve, Rapid desynchronization of an electrically coupled interneuron network with sparse excitatory synaptic input, *Neuron*, 67, 435–451, 2010.

27. M. Sahores and A. M. Naranjo, Gap junctions as therapeutic targets in brain injury following hypoxia-ischemia, *Recent Pat. CNS Drug Discov.*, 3, 209–215, 2008.

28. M. Galarreta and S. Hestrin, Electrical synapses between GABA-releasing interneurons, *Nat. Rev. Neurosci.*, 2, 425–433, 2001.

29. S. Jacquir, S. Binczak, J. M. Bilbault, V. Kazantsev, and V. Nekorkin, Synaptic coupling between two electronic neurons, *Nonlinear Dyn.*, 44, 29–36, 2006.

30. J. G. Mancilla, T. J. Lewis, D. J. Pinto, J. Rinzel, and B. W. Connors, Synchronization of electrically coupled pairs of inhibitory interneurons in neocortex, *J Neurosci.*, 27, 2058–2073, 2007.

31. F. Postma, C. H. Liu, C. Dietsche, M. Khan, H. K. Lee, D. Paul, and P. O. Kanold, Electrical synapses formed by connexin36 regulate inhibition- and experience-dependent plasticity, *PNAS*, 108, 13770–13775, 2011.

32. D. H. Pozza, C. S. Potes, P. A. Barroso, L. Azevedo, J. M. Lopes, and F. L. Neto, Nociceptive behaviour upon modulation of mu-opioid receptors in the ventrobasal complex of the thalamus of rats, *Pain*, 148, 492–502, 2010.

33. P. R. Parker, S. J. Cruikshank, and B. W. Connors, Stability of electrical coupling despite massive developmental changes of intrinsic neuronal physiology, *J Neurosci.*, 29, 9761–9770, 2009.

34. J. G. Mancilla, T. J. Lewis, D. J. Pinto, J. Rinzel, and B. W. Connors, Synchronization of electrically coupled pairs of inhibitory interneurons in neocortex, *J Neurosci.*, 27, 2058–2073, 2007.

35. N. Thammawongsa, N. Moongfangklang, S. Mitatha, and P. P. Yupapin, Novel nanoantenna system design using photonic spin in a PANDA ring resonator, *PIER Lett.*, 31, 75–87, 2012.

36. A. van der Horst and N. R. Forde, Calibration of dynamic holographic optical tweezers for force measurements on biomaterials, *Opt. Express*, 16, 20987–21003, 2008.

37. S. Pangratz-Fuehrer and S. Hestrin, Synaptogenesis of electrical and GABAergic synapses of fast-spiking inhibitory neurons in the neocortex, *J. Neurosci.*, 31, 10767–10775, 2011.
38. I. Srithanachai, S. Ueamanapong, S. Niemcharoen, and P. P. Yupapin, Novel design of solar cell efficiency improvement using an embedded electron accelerator on-chip, *Opt. Express,* 20, 12640–12648, 2012.

14 Future Challenges

14.1 INTRODUCTION

Whispering gallery modes (WGMs) of waves in nature such as electromagnetic wave, sound wave, and matter wave have shown interesting results, which can be useful for fundamental studies and applications in optoelectronics and nanoelectronics, especially after the announcement of the 2012 Nobel Prize in Physics for WGMs [1,2], where the authors confirmed that atoms can be trapped (stopped) by using WGMs in a microsphere. However, there are two more kinds of devices that can be used to trap light beams (atoms); the use of microcavity arrays performed by Yanik and Fan [3], and a nonlinear microring resonator by Yupapin and Pornsuwancharoen [4] for stopping light. Recently, Ang and Ngo [5] have also done experiments with slowing light in microresonators using a microring system. In this article, a new design of microring resonator device is proposed, which can be used to generate four forms of light simultaneously on a chip, whereas the storing and harvesting of trapped atoms/molecules can also be available. The proposed device is made up of silica and InGaAsP/InP with linear optical add-drop filter incorporating two non-linear micro/nano rings on both sides of the center ring (modified add-drop filter). This particular configuration is known as a PANDA ring resonator [6] as shown in Figure 14.1. Light pulse, for instance, Gaussian, and bright and dark solitons are fed into the system through different ports such as an add port and through port. By using the practical device parameters, the simulation results are obtained using the Optiwave and MATLAB programs. Results obtained by both analytical and numerical methods show that many applications can be exploited. In application, when the practical device parameters are used, then such a device can be fabricated and implemented in the near future.

14.2 RESULTS

For the whispering gallery mode, the result is obtained using the Optiwave program as shown in Figure 14.2. The ring material is InGaAsP/InP, where the device parameters are given in the caption of Figure 14.2. By using the MATLAB program, the whispering gallery modes of four state of light, that is, fast, slow, stopping, and storing can be generated and controlled simultaneously on-chip as shown in Figure 14.3. The storing stage can be seen easily, while the stopping condition can be observed and the following conditions are satisfied: (i) the center signal is lost in time between fast and slow signals or (ii) there is no movement among trapped particles or molecules, that is, the exchange of angular momentum introduces the conservation of angular momentum, where the combination of scattering and gradient forces is balanced under the adiabatic process.

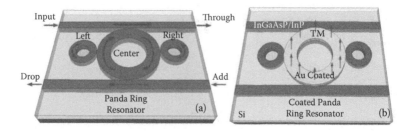

FIGURE 14.1 (a) A conventional PANDA ring planar waveguide was named and designed by Uomwech et al. (From K. Uomwech, K. Sarapat, and P. P. Yupapin, Dynamic Modulated Gaussian Pulse Propagation within the Double PANDA Ring Resonator System, *Microw. & Opt. Techn. Lett.*, 52, 1818–1821, 2010.) (b) Gold coated PANDA ring resonator for TM polarized coupling device.

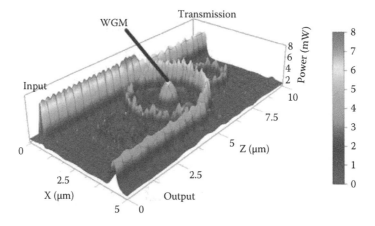

FIGURE 14.2 Result of whispering gallery mode of light within a PANDA ring waveguide InGaAsP/InP, $R_1 = R_2 = 0.775$ μm, $R_{ad} = 1.565$, $A_{eff} = 0.3$ μm², $n_{eff} = 3.14$, $n_2 = 1.3 \times 10^{-3}$ cm²/W, $\kappa_1 = \kappa_2 = \kappa_3 = \kappa_4 = 0.5$, $\gamma = 0.01$, $\lambda_0 = 1550$ nm.

The stopping light in term of signal condition can be easily performed using the whispering gallery mode concept, where the fast and slow light can be used as the upper and lower time frames or upper side and lower side peak signals for storing light at the center as shown in Figure 14.3, where in this case the movement (modulated signals) longer than 150 fs, that is, ms, ns, and ps are observed (stopped). The input pulse is a Gaussian pulse with pulse width of 100 fs, where the fast and slow time interval is known, however, the whispering gallery modes can be seen only under the resonant condition.

The use of a light-trapping probe for atom/molecule trapping and transportation (dynamically trapping) can be formed with a wide range of applications. In this case, the modulated signal is required to switch off the whispering gallery mode power via the add port, where atoms/molecules at the device center can be trapped and transported along the waveguide by the surface plasmon tweezers as shown in

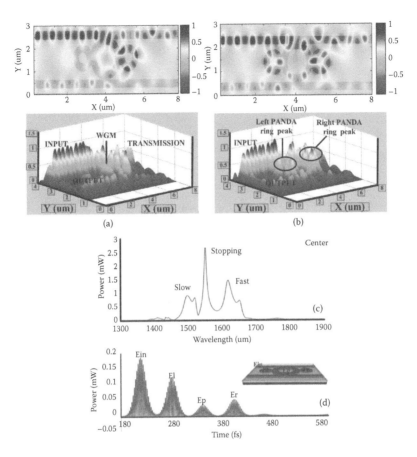

FIGURE 14.3 Stopping and storing light simultaneously detected using a PANDA ring, where (a) Center ring, (b) Side rings, (c) Center peak and side peaks, and (d) Fast (2nd peak) and slow (4th peak) light detected at drop port with time interval of 150 fs.

Figure 14.4. The dynamic tweezers are generated by a PANDA ring, whereas in practice, particle angular momentum can be introduced by a metal coating material on the waveguide surface or combining the external modulation via the add port, which can be used to trap and transport the atom/molecule to the required destination when the gradient force is greater than the scattering force along the waveguide. The use of such a concept for a new type of solar cells is shown in Figure 14.5, where free electrons from the depletion region can be trapped and transported (injected) to the metal contact faster than the conventional device; the solar cells efficiency can be increased by five times to the conventional one [7].

For the above reasons, we found that the WGM of light can be easily formed by a microring resonator [8] or nonlinear coupling effects to the center ring of a PANDA ring [9]. In Figure 14.6, the WGM and leaky modes are generated by a PANDA ring, which can be formed in the same way for large-scale motion. Thus, the use of WGM for large-scale motion can be useful, for instance, the WGM

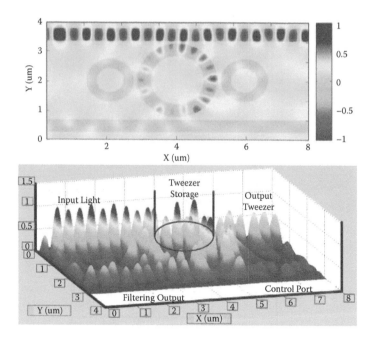

FIGURE 14.4 Dynamic tweezers are generated by a PANDA ring and transmitted via a through port for atom/molecule harvesting and transportation.

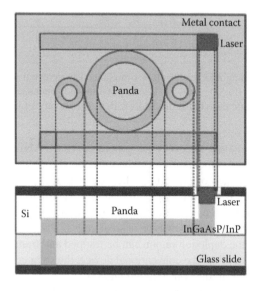

FIGURE 14.5 New type of solar cells using a PANDA and embedded particle accelerator (optical trapping and transportation) (From N. Thammawongsa et al., Storing and Harvesting Atoms/Molecules On-Chip: Challenges and Applications. *J. Biosensors & Bioelectronics*, 3, e114–115, 2012; I. Srithanachai et al., Novel Design of Solar Cell Efficiency Improvement Using an Embedded Electron Accelerator On-Chip, *Opt. Exp.*, 20, 12640–12648, 2012.)

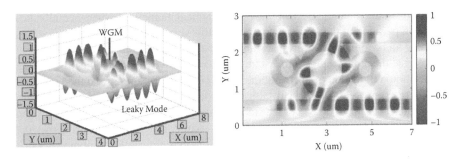

FIGURE 14.6   WGMs and leaky modes generated by a PANDA ring.

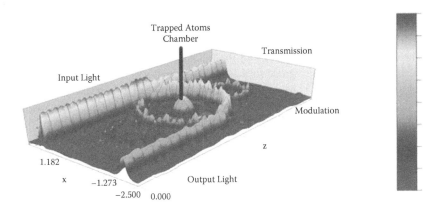

FIGURE 14.7   Schematic diagram of trapping and modulating atoms by light.

between the Earth and Moon orbital motion, where in this case the matter wave (wave-particle duality) can be formed by the Moon's orbital motion around the Earth, which can be useful for Earth disaster investigation and prediction. Another application is atomic modulation, which can be formed by using the WGM of light in a PANDA ring as shown in Figure 14.7, where the atoms are dancing and atomic radio can be performed and realized by the external modulation to the trapped atoms [10].

14.3   METHODS

To perform the experiment in which atoms can interact with light naturally requires a clean source of atoms, where one of the techniques uses a standard magneto-optic trap (MOT) [11], in which analysis of the cold collision measurements is performed in a high-gradient magneto-optical trap with a few trapped cesium (Cs) atoms. The net result is a cloud of cold, confined cesium atoms inside a vacuum chamber, providing a convenient initial condition for optical-lattice experiments. While in the optical lattice, the trapping fields are extinguished so that they do not interfere with the atomic dynamics in the lattice.

The primary challenge in observing dynamical tunneling experimentally lies in the preparation of the initial state. In brief, this state-preparation process involves trapping and cooling the atoms in a standard magneto-optic trap, applying laser light to polarize the atomic spins into a state that is insensitive to magnetic fields, driving a two-photon transition to select out the coldest 0.3% of the atoms, and then manipulating the atoms with the optical lattice to produce a nearly minimum uncertainty wave packet localized on one of the stability islands. Small-scale optical devices can be used as an electron correlation generation source, in which electrons across the gap are driven by quantum tunneling and require a new description of nonlocal transport, which is crucial in nanoscale optoelectronics and single-molecule electronics.

In our finite difference time domain (FDTD) simulation, the perfectly matched layer (PML) absorbing boundary conditions applied by the Berenger [12] and Yee scheme [13] absorb the electromagnetic wave without any reflection at the computational boundary. A 100 fs-Gaussian pulse modulated by a 200 THz carrier is exited. The vertical waveguide thickness and material composition is accounted for by computing the effective refractive index n_{eff} for the fundamental mode at $\lambda = 1.55$ μm. In the vertical direction, each waveguide structure is 0.45 μm thick, vertical core thicknesses of 0.3 μm to 0.5 μm, n_{eff} is between 3.2 to 3.4, in which the parameters are obtained by using the practical material parameters of InGaAsP/InP. Therefore, the waveguide core n = 3.14 is bordered on each side by air n = 1. The parameters for add-drop optical multiplexer and both nanorings on the left- and right-hand sides of the PANDA ring are set at $R_1 = R_r = 0.75$ μm and radius of the center ring is $R_{ad} = 1.5$ μm. The coupling coefficient ratios are $\kappa_1 = \kappa_4 = 0.5$, $\kappa_2 = \kappa_3 = 0.4$, effective core area of the waveguide is $A_{eff} = 0.25$ μm², and waveguide loss coefficient is $\alpha = 0.1$ dB/mm. When gold material is coupled as a function of frequency by coating on the waveguide as shown in Figure 14.1, the free-carrier concentration of gold is $N \approx 6 \times 10^{22}$ cm^{-3} and the plasma frequency is $\omega_p/2\pi \approx 2 \times 10^3$.

In this work, the whispering gallery modes can be generated by trapping particle/photons within PANDA rings as shown in Figure 14.8, where the tunneling

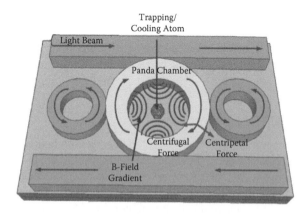

FIGURE 14.8 PANDA chamber for particle/photon trapping and storage under (i) centripetal and centrifugal forces and (ii) WGMs.

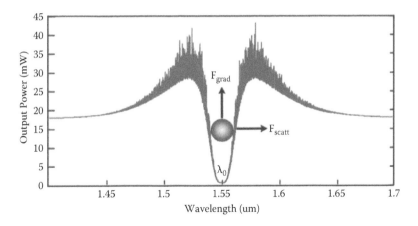

FIGURE 14.9   Particle (photon) is trapped by a potential well within a PANDA ring wave-guide, where F_{grad}: gradient force, F_{scatt}: scattering force.

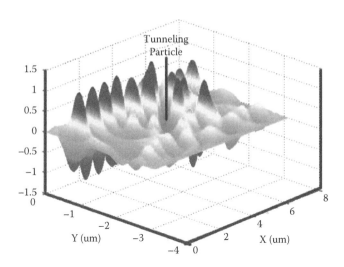

FIGURE 14.10   Dynamic tunneling particles/photons, where X and Y: two-dimensional scale.

particles/photons can be generated when the particle energy is greater than the trapping potential as shown in Figure 14.9. By using the MATLAB program, the whispering gallery modes of tunneling particles/photons can be generated and controlled on-chip as shown in Figure 14.10. The tunneling particles/photons can be confined by the WGMs at the center, which can be used as a particle source for various applications as shown in Figure 14.11.

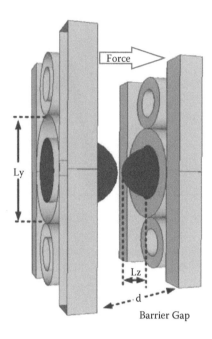

FIGURE 14.11 WGMs switching control for particle/photon confinement (storage) and harvesting, where Ly, Lz: WGM's dimension; d: device (barrier) gap.

14.4 CONCLUSION

The proposed system can be used for applications such as atom/molecule transportation, everlasting atom/molecule investigation, atomic/molecular storage for quantum gate or computer application, storm search and navigator sensors, microplasma source, 3-D flat panel device and large cooling area (volume), and so forth. In this work, one of them is demonstrated. Under the stopping and storing condition, the system is considered as the storage unit, in which the atom or molecule can be trapped by the whispering gallery light beam at the center, in which the trapped atoms/molecules can be modulated by the external modulated signals. Forms of light in a PANDA ring resonator with or without coated material can be manipulated, and it can be observed that the two nonlinear side rings have shown interesting results and aspects. The input light can be in the forms of soliton or Gaussian pulses. The use of photon or matter wave as input is also possible. Particularly, an interesting aspect can also be formed by using the trapped electron, in which the matter wave concept can arise within the PANDA ring waveguide. In this work, we found that four behaviors of light, for instance, fast, slow, stopping, and storing can be manipulated and seen simultaneously by using the PANDA ring planar waveguide, which can be fabricated and tested on-chip. The expected output light can be in the forms of surface plasmon, potential wells, leaky modes, whispering gallery modes, matter wave, and photons (particles). The use of nonlinear Schrödinger's equation can also be available, where in this case the propagation of light is treated as a particle (photon) within the PANDA ring, in which the tunneling effects of particles can be performed and investigated.

REFERENCES

1. D. J. Wineland, J. J. Bollinger, W. M. Itano, and J. D. Prestage, Angular momentum of trapped atomic particles, *JOSA B*, 2, 1721–1730, 1985.
2. J. C. Knight et al., Characterizing whispering-gallery modes in microspheres by direct observation of the optical standing-wave pattern in the near field, *Opt. Lett.*, 21, 698–670, 1996.
3. M. F. Yanik and S. Fan, Stopping and storing light coherently, *Phys. Rev. Lett.*, 92, 083901–3, 2004.
4. P. P. Yupapin and N. Pornsuwancharoen, Proposed nonlinear microring resonator arrangement for stopping and storing light, *IEEE Photon. Techn. Lett.*, 21, 404–406, 2009.
5. T. Y. L. Ang and N. Q. Ngo, Tunable flat-band slow light via contra-propagating cavity modes in twin-coupled microresonators, *JOSA B*, 29, 924–933, 2012.
6. K. Uomwech, K. Sarapat, and P. P. Yupapin, Dynamic modulated Gaussian pulse propagation within the double PANDA ring resonator system, *Microw. & Opt. Techn. Lett.*, 52, 1818–1821, 2010.
7. I. Srithanachai, S. Ueamanapong, S. Niemcharoen, and P. P. Yupapin, Novel design of solar cell efficiency improvement using an embedded electron accelerator on-chip, *Opt. Exp.*, 20, 12640–12648, 2012.
8. S. Kamoldilok and P. P. Yupapin, Nano heat source generated by leaky light mode within a nano-waveguide for small electrical power generator, *Ener. Conver. & Manag.* 64, 23–27, 2012.
9. N. Thammawongsa, S. Tunsiri, M. A. Jalil, J. Ali, and P. P. Yupapin, Storing and harvesting atoms/molecules on-chip: Challenges and applications, *J. Biosensors & Bioelectronics*, 3, e114–115, 2012.
10. N. Thammawongsa, N. Moonfangklang, S. Mitatha, and P. P. Yupapin, Novel nano-antenna system design using photonics spin in a panda ring resonator, *PIER L,* 3, 75–87, 2012.
11. B. Ueberholz, S. Kuhr, D. Frese, V. Gomer, and D. Meschede, Cold collisions in a high-gradient magneto-optical trap, *J. Phys. B: At. Mol. Opt. Phys.*, 35, 4899–4914, 2002.
12. J. P. Berenger, Perfectly matched layer for the FDTD solution of wave structure interaction problem, *IEEE Trans. Antennas Propagation,* 44, 110–118, 1996.
13. K. S. Yee, Numerical solution of initial boundary value problems involving Maxwell's equations in isotropic media, *IEEE Trans. Antennas Propagation*, 14, 302–307, 1966.

Index